Practical Action Publishing Ltd
27a Albert Street, Rugby, CV21 2SG, Warwickshire, UK
www.practicalactionpublishing.org

in association with

The Water, Engineering and Development Centre (WEDC)
Loughborough University of Technology
Leicestershire LE11 3TU, UK

© WEDC, 1994

First published in 1995
Transferred to digital printing in 2008

ISBN 978 1 85339 294 8

Since 1974, Practical Action Publishing has published and disseminated books and information in support of international development work throughout the world. Practical Action Publishing (formerly ITDG Publishing) is a trading name of Intermediate Technology Publications Ltd (Company Reg. No. 1159018), the wholly owned publishing company of Intermediate Technology Development Group Ltd (working name Practical Action). Practical Action Publishing trades only in support of its parent charity objectives and any profits are covenanted back to Practical Action (Charity Reg. No. 247257, Group VAT Registration No. 880 9924 76).

Conference Secretary: Rowena Steele
Papers produced by Karen Betts
Cover design: Rod Shaw
Cover photograph: Audio Visual Services, Loughborough University of Technology

AFFORDABLE WATER SUPPLY AND SANITATION

AFFORDABLE WATER SUPPLY AND SANITATION

Selected papers of the 20th WEDC Conference
Colombo, Sri Lanka, 1994

Edited by John Pickford,
Peter Barker, Adrian Coad, Tom Dijkstra,
Bob Elson, Margaret Ince and Rod Shaw

Intermediate Technology Publications
in association with
The Water, Engineering and Development Centre
1995

IT Publications, 103/105 Southampton Row, London WC1B 4HH, UK.
WEDC, Loughborough University of Technology, Leicestershire, LE11 3TU, UK.

A CIP record for this book is available from the British Library.

ISBN 1 85339 294 4

Conference Secretary: Rowena Steele
Papers produced by Karen Betts
Cover design: Rod Shaw
Cover photograph: Audio Visual Services, Loughborough University of Technology

Printed by Blenfield Printers Ltd., Leicester, UK.

ACKNOWLEDGEMENTS

The editors would like to acknowledge the commitment
of the Local Organizing Committee whose arrangements helped to ensure
the success of the 20th WEDC Conference.

CONTENTS

INTRODUCTION

by John Pickford

The venues for WEDC Conferences alternate between Africa and Asia, with many good locations. Kathmandu in 1992 and Accra in 1993 provided excellent surroundings, but nothing compares with the delightful surroundings and facilities of the Taj Samudra Hotel in Sri Lanka. This was the setting for the 20th WEDC Conference held in August 1994. The Conference Centre is situated on high land overlooking the Atlantic Ocean on one side and the City of Colombo on the other. Warm sea breezes, friendly and helpful staff, and fine catering completed the delight for participants.

Over a hundred papers were presented and by holding most sessions with four simultaneous groups there was plenty of time for discussion. Participants took advantage of this by sharing their experience, which covered great expanses of ground both geographically and in terms of professional discipline.

For this volume we have selected forty-three papers from the more-than-a-hundred presented. Sixty per cent of selected authors were from South Asia — the area formerly known as the Indian sub-continent. Other authors came from a variety of places — three from Africa, from Australia, Japan, the Philippines, and several (including WEDC staff) from Europe.

The selected papers represent the range of topics that is usual at WEDC Conferences. Many deal with 'software' aspects — people, communities, health, management and institutions. On the technological side, water supply projects attracted most attention, but there was a greater-than-usual prominence for solid waste management, that mirrors increasing world-wide concern with rubbish.

The papers in this volume also represent the principles for which WEDC is well-known and which WEDC Conferences emphasize. The theme of the Sri Lanka meeting was *affordable* water supply and sanitation. Affordability underlies all the papers, rather than receiving specific mention in the text.

WEDC is concerned with improving infrastructure in low-income and middle-income countries. So the technology and software which the Conference discussed was inevitably that which is appropriate for and affordable by low-income and middle-income people.

SECTION 1

COMMUNITIES, HEALTH, PEOPLE

Mapping of fluorosis affected villages

B.V. Apparao, Mrs Anitha Pius and G. Karthikeyan, Tami Nadu, South India

EXCESS FLUORIDE IN drinking water sources beyond a tolerance limit is responsible for the disease 'fluorosis', a serious public health problem in several parts of the world. Identification and mapping of fluorosis affected villages is the first step in the direction of mitigating and controlling the problem. A systematic study has been carried out in order to map the fluorotic villages of Dindigul Anna district of Tamil Nadu, South India and a suitable methodology has been developed for the purpose. This methodology for mapping the fluorosis affected villages is presented in this paper.

Survey of the villages

Survey of the school children (8-16 years age group) in order to examine the symptoms of dental fluorosis (cf. fig 1) is the first and foremost step which can decide the prevalence or absence of fluorosis.

Only when prevalence of dental fluorosis is confirmed in this survey, is it necessary to study further the magnitude and severity of the problem. Door to door survey of each and every family in the village for examination of the symptoms of dental fluorosis and skeletal fluorosis (cf. fig. 2) is the next step which gives us the percentage incidence of fluorosis as well as community fluorosis index.

Community fluorosis index (C.F.I) can be calculated as follows (Dean, H.T. and Elvove, E, 1935). Based on the symptoms dental fluorosis is classified into seven categories, viz., normal, questionable, very mild, mild, moderate, moderately severe and severe and each of these seven

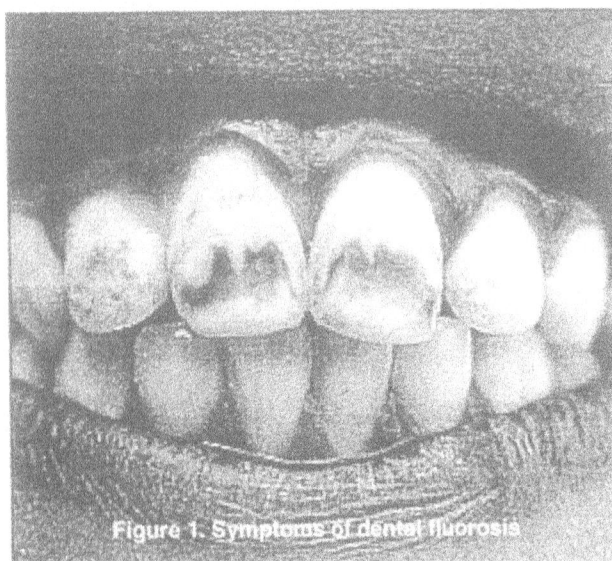

Figure 2. Symptoms of skeletal fluorosis

classifications is given a numerical weight such as 0, 0.5, 1, 1.5, 2, 3 and 4 respectively. During door to door survey, people are classified into various categories as per the above classification. The number of people in each category is multiplied by the corresponding numerical weight, the products thus obtained for the various categories are added up and the sum total divided by the total number of people surveyed, gives the community fluorosis index. Only when the C.F.I. value is greater than 0.6, is fluorosis considered to be a public health problem in that area. The results of percentage incidence of fluorosis and community fluorosis index obtained for different areas based on this methodology are given in tables 1, 2 and 3 respectively.

Water analysis for fluorides

For those villages which have C.F.I.>0.6, all the available water sources are analysed for fluoride by the ion-selec-

Figure 1. Symptoms of dental fluorosis

Table 1. % incidence of fluorosis among children

Area	No. of children surveyed	% incidenc of fluorosis	
		dental	skeletal
Control	420	0	0
Fluorotic area 1	412	30	0
Fluorotic area 2	410	58	0
Fluorotic area 3	420	89	0

Table 2. % incidence of fluorosis among adults

Area	No. of adults surveyed	% incidenc of fluorosis	
		dental	skeletal
Control	310	0	0
Fluorotic area 1	308	39	0
Fluorotic area 2	312	64	0
Fluorotic area 3	314	93	34

Table 3. Community fluorosis index data

Area	No. of subjects surveyed	Community fluorosis index
Control	730	0
Fluorotic area 1	720	0.83
Fluorotic area 2	722	1.85
Fluorotic area 3	734	3.72

tive electrode method (Fluoride Electrode Instruction Manual, 1977). In the present study fluoride was estimated using an expandible ion analyser EA 920, the fluoride ion selective electrode 9409 and the reference electrode 90-00-01 (all Orion, USA make). Total ionic strength adjustment buffer made from cyclohexylene dinitrilotetracetic acid (CDTA), acetic acid and sodium chloride, is added to the standard fluoride solutions as well as the samples before measurement of fluoride. The instrument is calibrated with two standard solutions so chosen that the concentration of one is 10 times the concentration of the other and also that the concentration of the unknown falls between those two standards. Then, the concentration of the unknown is directly read from the digital display of the meter.

Mapping the fluorotic area

Based on the results of percentage incidence of fluorosis, community fluorosis index and the fluoride levels of drinking water sources for each village, fluorosis maps for each and every block of a district are prepared. The fluorosis map of a block shows the absence/prevalence of fluorosis in any village of that block. It also indicates the priority village panchayats where drinking water sources contain fluoride greater than 1.5 pm, for immediate attention of the Government to provide alternate safe drinking water supply. From the fluorosis map one can know even the range of fluorides in the drinking water sources of the Priority Panchayats. Fluorosis maps have been prepared for all the blocks of the Dindigul Anna District and these are found to be quite useful by the Tamil Nadu Water and Drainage Board, Government of Tamil Nadu in order to implement the scheme of provision of safe drinking water supply to rural areas on a priority basis.

References

Dean, H.T. and Elvove, E., 'Public Health Report', Washington, 50(1935) 1719.

'Fluoride Electrode Instruction Manual', Orion Research Inc., U.S.A. (1977).

Sarvodya's integrated approach to water and sanitation

Vinya S. Ariyaratne and Palitha Jayaweera, Sri Lanka.

THE SARVODAYA SHRAMADANA MOVEMENT (SSM) of Sri Lanka is by far the largest non-governmental organization (NGO) in the country. It is also the largest NGO working in water and sanitation sector in Sri Lanka. The methods and techniques developed by Sarvodaya in building community awareness, ensuring community participation and sustaining community management of integrated community water supply and sanitation schemes in hundreds of Sri Lanka villages on a national scale has been its greatest achievement.

Based on its extensive field experience Sarvodaya has postulated a five-stage model of village community development (Figure 1) and, water and sanitation related activities are initiated in the Third Stage of village development. The *First Stage* is that of psychological infrastructure building. It begins quite simply with a village level discussion about local needs and organizing self-help activities. Villages enter the *Second Stage* of social infrastructure building when they have formed one or more community groups of; farmers, mothers, children, youth and elders. The *Third Stage* of the Sarvodaya development process is a very critical one. At this stage the village is organized to satisfy its own basic and secondary needs. It is at this stage that water and sanitation related programmes are initiated. In addition the village Sarvodaya groups are brought together and institutionalized as a legally incorporated body (the Sarvodaya Shramadana Society) which is entitled to open its own bank account, obtain loans and start economic activities with support from District level and National Level Sarvodaya structures. Villages in the *Fourth Stage* are expected to become self financing in their Sarvodaya activity and they assist neighbouring villages as well in their *Fifth Stage*.

The Sarvodaya Rural Technical Services (SRTS) which is one of the eight main programmes of the Sarvodaya Shramadana Movement, is responsible for implementing integrated community water supply and sanitation projects in 16 out of 25 administrative districts in Sri Lanka. It provides technical and financial assistance to village societies in relation to meeting their basic needs primarily with respect to water supply, sanitation, energy and transport. Community water and sanitation needs are satisfied through the construction of gravity water supply schemes, hand dug shallow wells and low-cost latrines, with maximum community participation and mobilization of local resources.

Once the legally independent Sarvodaya Shramadana Society (SSS) is formed at the Third Stage of village development, the society can own land, buildings, vehicles, equipment, and receive loans and seed capital from any external agency other than Sarvodaya itself.

A close contact is maintained between these societies by a full-time Sarvodaya worker known as the Gramadana Worker. Gramadana is a word which means sharing of resources at the level of villagers on a voluntary basis. Several clusters of such villagers are organized around Sarvodaya Divisional Centres, and several divisional centres around District Centres which correspond to the government administrative units.

The decision-making process involved in a typical SRTS water project is presented in Figure 2. Firstly, a formal request seeking assistance for the need for improved water and sanitaion is submitted to the Sarvodaya District Coordinator by the officials of the SSS through the Gramadana Worker. On receiving such a request the District Coordinator will assess the availability of funding and hand it over to the SRTS technician for further action. The technician will visit the village with the Gramadana Worker and the members of the society and have a preliminary participatory survey to check the technical feasibilty of the project. This survey will lead to the discussions with the SSS to decide the technology and options which will be selected for the project. If the project is technically feasible SSS will start the mobilization and educational activities with the assistance of the Gramadana Worker. In the mean time SRTS technician will prepare the necessary plans and the cost estimates for the approval of the Sarvodaya Head Office. The financial input provided by Sarvodaya will be complimentary to whatever the villagers themselves can provide within their means. This would ensure village self reliance and self confidence. All critical decisions as to the choice of technology, levels of service, location of facilities etc., are taken by the community, in additon to their cash, material and labour contributions.

As soon as the Head Office approves the project, the funds would be released to enable the SRTS technician to start the construction work on a shramadana basis. With the commencement of the construction work, the SRTS technician will train two village youths who have been selected by the SSS as Caretakers for the maintenance of the project. While SRTS contributes financially and technically, the society will provide the available raw materials at village level and the unskilled labour required. On completion of the project, the scheme operation and maintenance (O & M) will be formally taken over by the society and trained caretakers will look after the project under the guidance of the SSS. O&M will be the sole

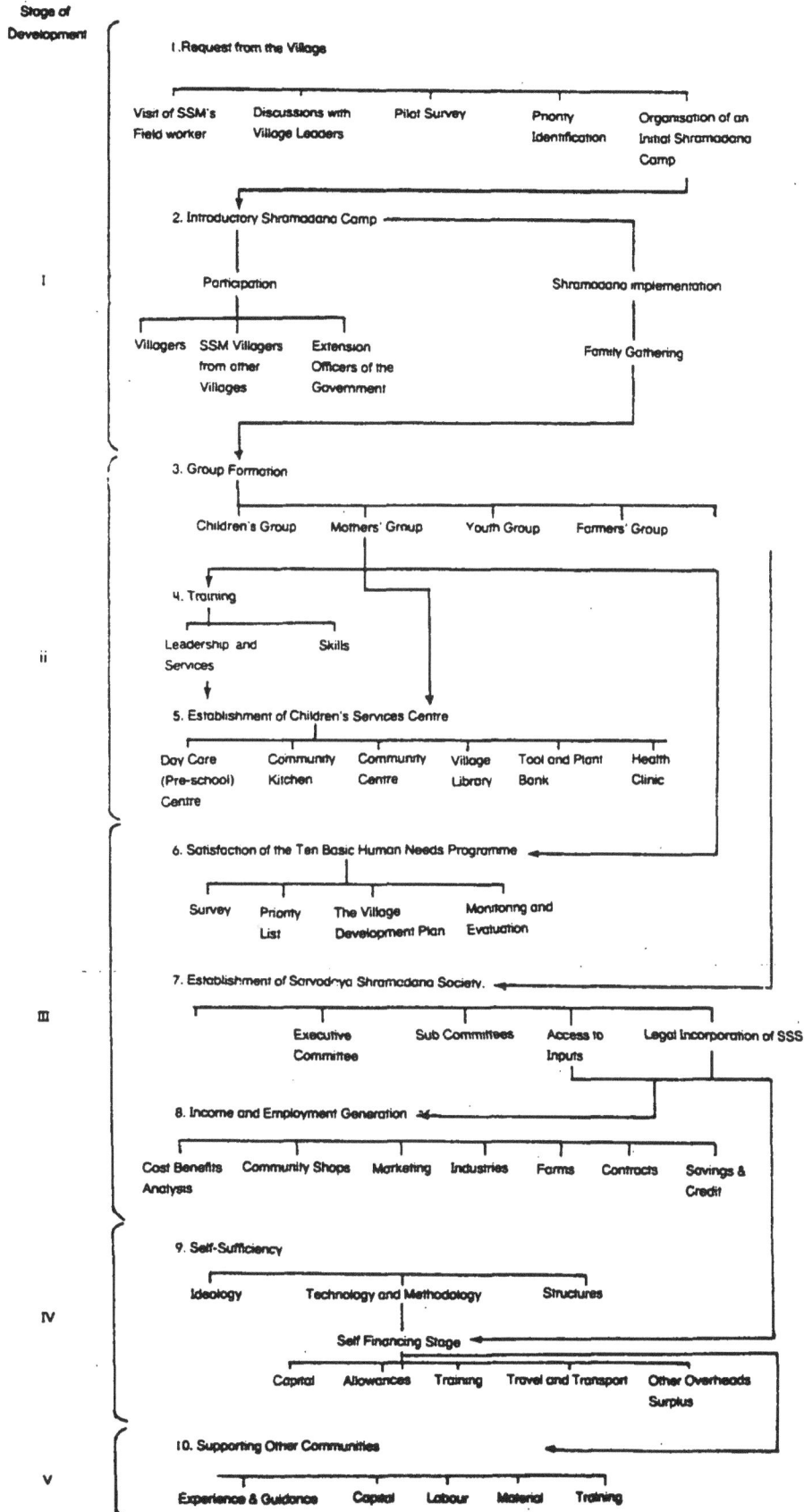

Figure 1. Sarvodaya Five-Stage Village Development Model.

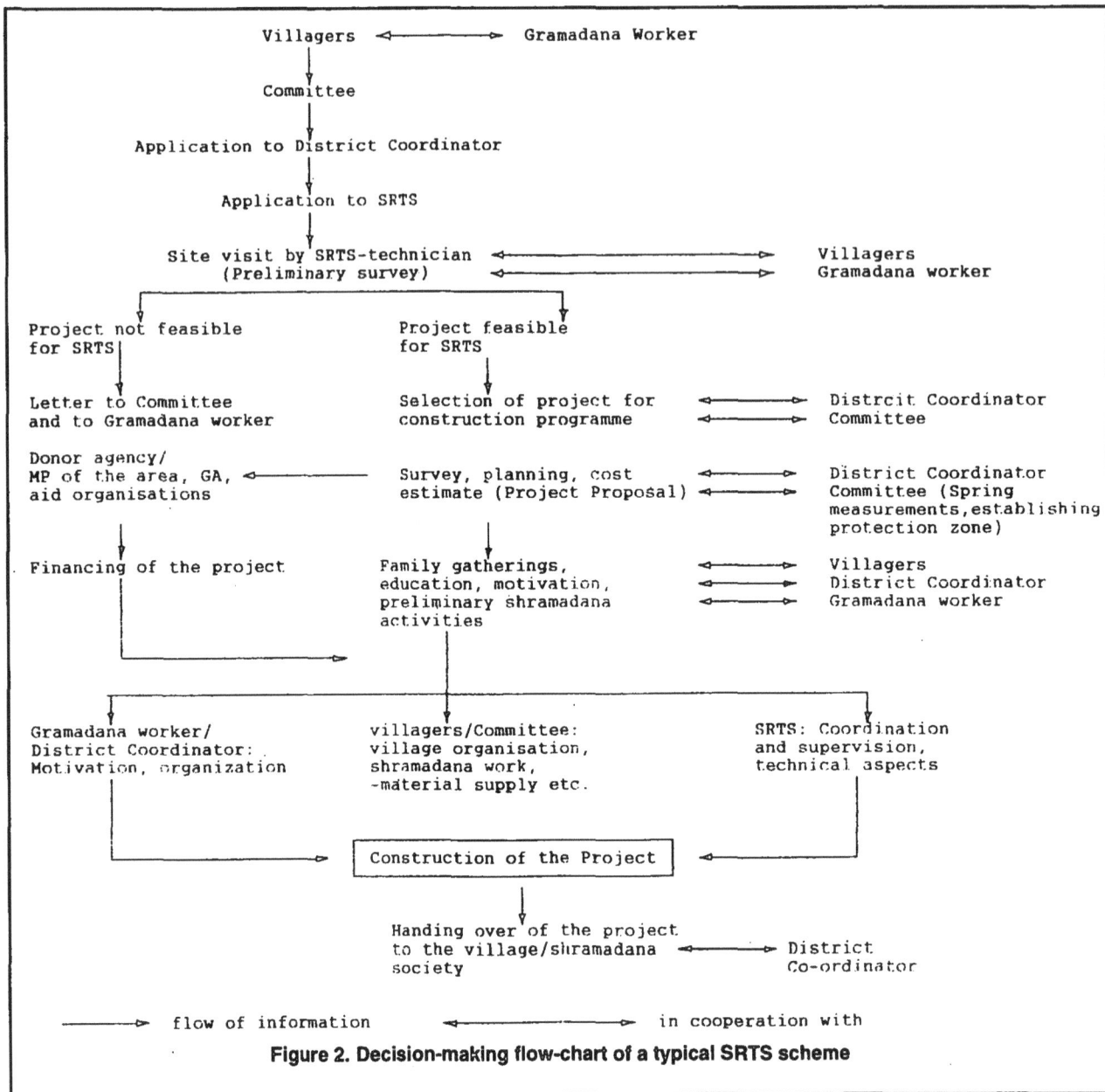

Figure 2. Decision-making flow-chart of a typical SRTS scheme

responsibility of the SSS but if there is any technical problem which goes beyond the capability of the caretakers the SSS can still contact the SRTS technicians for assistance.

During construction phase, an accompanying hygiene education activity is implemented through the pre-school and other social programmes initiated by Sarvodaya in the village. Women are involved through mothers' groups and are represented in the village SSS. Thus, Sarvodaya's holistic approach addresses both technical and social issues related to community water supply and sanitation while the activities that are carried out during the succeeding stages of village development directly address the root causes of ill-health; poverty and powerlessness. Further, Sarvodaya has successfully demonstrated that water and sanitation needs could be satisfied through an integrated community development approach which recognizes amongst others, the spiritual values cherished by the people.

Gender appreciation in Gujarat, India

J.M. Barot, Gujarat Jalseva Training Institute, Gujarat, India

WATER SUPPLY AND SANITATION services are the basic necessities of the community and are a pre-condition for development. They also play an important role in improving the health and quality of life.It is therefore rightly said that *efficient water and sanitation services are the indicators of progress made by a nation.*

Gujarat is one of the states in India which has put in commandable efforts to provide safe water supply to its 41.2 million people, 66 per cent of which are living in rural areas. Out of a total of 18 569 villages in the state, as many as 14 503 villages were having either quality or quantity problems of water and were declared as *No Source* villages. The State Government in support of Union Government, the Royal Netherlands Government and the World Bank, has so far (March 1994) covered 14 407 villages by providing safe and dependable source of water with delivery systems.

However to achieve the set objectives, it is essential to have sustainability of the systems which are developed and also provide satisfaction to the consumers. Obviously, this can not be accomplished without the support and involvement of community for whom the services are created. Since women are main caretakers of water and sanitation facilities, their role and participation assumes great significance.

The State Government through its Gujarat Water Supply and Sewerage Board (GWSSB) and other departments, has initiated various actions to appreciate and encourage involvement and participation of women in this vital sector and the results are quite encouraging.

Why focus on women ?

Looking to the Indian conditions and socio-cultural traditions, women have to pay a major role in the management of water and sanitary services at domestic level. They play many roles in the society, as a mother, housewife, cook, cattle breeder and artisan. Traditionally, they are the custodians and managers of water in the house and play a role in management of water sources and environmental hygiene in their neighbourhood. They draw, store, utilize and manage water as per requirements of the family. Many a time girls are also engaged in fetching water.

So far as health of the people is concerned, women provide most health care. In a family they prepare food and nourish children, they clean the house and sweep the yard, they dispose of children's excreta and other domestic wastes including cattle waste. They teach hygienic habits to the children and care for the sick in the family.

Almost all traditional birth attendants and nurses are women. Thus programmes on health and sanitation can not succeed without the active role played by the women as they better understand their own problem and can solve them.

Women are considered to be best teachers and have potential influence on the family members and particularly children. They have also got a high degree of tolerance and capacity to work. So their involvement can be beneficial in three ways, as a beneficiary, as a mother or a family member and as a motivator.

Out of the total 41.2 million population of Gujarat, nearly half i.e. 19.9 million is female, of which 65.5 per cent are living in villages. It is therefore essential to join the other half of the society in developmental activities and especially those activities concerning them the most. Literacy rate of women particularly in rural areas is less, only 33 per cent and they are poorly represented. In such circumstances no progress or success could be achieved without recognizing the importance and role of women who are considered to be *agents for change.*

Considering above aspects and also the need to remove imbalance by uplifting their social status, womens' involvement in water and sanitation services can become an entry point for many other developmental activities.

Objectives for involvement of women

Since women are more concerned with water, for understanding their choices and conveniences, they need to be involved in the management of services right from the project stage and taken into confidence while selecting the source, location of structures and implementation of the project. Bringing water and sanitation facilities, nearer to their houses diminishes drudgery of women and saves considerable time which otherwise is spent in walking long distances. This leisure time could be fruitfully utilized for some productive work which can uplift their status, enhance their grip on life and their bid for betterment of their families.

Sustainability of the system which could be ensured through satisfactory operation and maintenance is most essential. Men go out of villages to work and other purposes but women are available in the villages for most of the time. So they need to be involved in operation and maintenance of the system. Doing so they will take care of equal distribution to all sections of the society, protect the assets and attend to minor repairing works e.g. handpump etc.

For utilizing their abilities and influence regarding health and hygiene aspects, women need to be awakened and motivated to participate in the health and sanitation programmes. This will help reduce the pollution and wastes of water resulting in reduction of water borne diseases.

For want of sanitation facilities in rural areas, women are facing lot of difficulties in defecation and taking bath. Avoiding nature's call till onset of darkness causes many health problems. Similarly non availability of bathing places forces them to avoid taking baths causing many skin diseases. Hence women need to be oriented towards the use of toilet and bathroom.

Current gender imbalance

The status of women is significantly low, resulting in many disadvantages and limitations to them. The present circumstances prevent them from participating freely and fully in the development programmes as well as improving their own life. Their activities and responsibilities are mostly confined to household only. These limitations remain as severe constraints while involving women in water and sanitation programmes.

The girls are usually engaged in household activities and are not sent to school for education. So the literacy rate among women is very low in rural areas. Without proper education, they cannot assimilate information so also, they cannot appreciate the importance of safe water and clean environment. The new technologies could also be unknown to them, as they remain the domain of men. Thus they are deprived of the technological developments and their use in reducing the drudgeries and sufferings.

Due to male-dominated society, women have less power and authority to take decisions. Decisions outside home are mostly taken by men. No properties are transferred on their name and no credits are given to women. This state of affairs deprives them of any authority or confidence to manage independently.

Although women are required to attend to sick persons in the family and provide health care, they have to rely on traditional knowledge and views on health and hygiene. They have no access to training facilities. Such training are limited to men only. In many cases, they themselves become the victim of unhygienic conditions and suffer miserably for want of knowledge or access to remedies.

Social customs and traditions forbid them to communicate with men members. Women are also not allowed to attend evening meetings or outstation visits. Similarly, socio-cultural traditions and prejudices have sex-specific beliefs. Due to menstruation, women and men are not allowed at many places to defecate and bath at the same place. Women have to go for defecation at night only for reasons of privacy, thus depriving them of the benefit of basic services.

Due to untouchability and other taboos, the women of backward communities are not allowed to use public sources of water supply and sanitation. Similarly, they do not attend meetings. And even if they attend the meetings, they do not speak out in open. Thus their difficulties remain untold and unsolved.

Findings of survey on water-related gender problems

The Royal Netherlands Government is supporting the State Government in providing safe water supply and sanitation facilities in acute problem areas. Three projects have been implemented with their financial support, in scarcity, salinity and fluoride prone areas.

In these areas, post-project implementation activities are also monitored by them in which support of Non-Governmental Organizations (NGO) is obtained. With a view to knowing the gender specific problems and for their redressal, a survey of 300 women in 30 villages was carried out (1989) with the support of an NGO (SEWA). The finding of the survey are as follows:

- 78 per cent of women spend about four hours daily fetching water.
- Women walk a minimum of six km to transport water.
- 53 per cent of women in drought prone areas complained that they could not bring enough water for even personal hygiene during menstruation and the post-natal period.
- 42 per cent of women replied that they were never consulted about the site of the bore well or the standpost of water supply.
- 63 per cent of women showed their readiness to learn skills of water harvesting, bunding, drip irrigation, afforestation and desalination.
- 86 per cent of women said that they did not enjoy seeing their children and family members getting sick from water-borne diseases due to dirty and contaminated water.
- In most of the projects, women are not consulted while deciding the site, budget, formulation of scheme and its operation/maintenance.

The report has also provided the following information.

- Due to migration of men to the cities in search of job, women have to take help from children in fetching water with the result that they miss school.
- Women living in desert areas have to travel long distances in summer in an attempt to save their cattle from heat. During the journey, they lose large numbers of cattle.
- Due to lack of safe and adequate water and sanitation facilities, many children die of diarrhoea and dehydration.
- Resort to unsafe source of water results in sickness, diseases, and death, especially of children.
- Due to the social practice of untouchability, women of scheduled caste/tribe communities do not have access to new sources and depend on traditional sources.

Strategies for a gender approach

To involve the women in the programme and motivate them to achieve targeted objectives, it is imperative to work out strategies and action plan for implementation. The first and foremost would be to recognize and accept the vital role to be played by women in the programme. Once their role is recognized, it is essential to create awareness among them about the issues most concerning them and support required from them.

The water supply and sanitation services are most beneficial to women as they ameliorate their drudgeries and health sufferings. The benefits available from such services need to be explained to them so as to seek their support in making the services sustainable. By explaining the scarcity of fresh water, women could be convinced to conserve water for its economic use. Similarly, the health implications of polluted water can convince them to protect the sources from pollution.

This leisure time could be utilized for some productive work and to train the family members especially children to use and keep the latrines and bathrooms clean. To bring women in to this process, special measures are needed to overcome the constrains described under Gender imbalance. Women need access to information and to take part in decision making and management of the village water system, not in name, but in reality. To do so, they need to be encouraged to attend project meetings either mixed, or with women alone, and choose their own representatives, in local water and health committees. The programme also needs to see that they are capacitated and get the scope to influence the sustainability of the water supply and sanitation facilities.

Regularizing operation and maintenance of services will not only create satisfaction among them but also create a sense of sharing responsibility. Maintenance and management of handpumps and village schemes can be done jointly by men and women, if both are capacitated. In regional schemes, management can only be done jointly, with the water authorities managing intake, transmission lines, treatment plant, etc. and the local communities managing the village distribution nets, and minor repairs.

Male members need to be sensitized so as to appreciate the partnership of women in the projects and encourage them. Providing back up support and funding by the implementing agencies will help women bring success in the programme. This process can also be accelerated by adopting an integrated approach with other NGOs and concerned departments.

Action taken

The State Government has made a modest beginning to involve women in the water supply and sanitation services by initiating various actions. The first thing to start with was the recognition and adoption of the fact that the role of women in this sector is vital and inevitable. The

authorities supported the Royal Netherlands Embassy (RNE) and a Gujarati NGO to carry out a fact finding survey on the status of service and problems of women.

The authorities also initiated dialogue with the NGOs to attract them in the field and motivate and enable village women to take interest. Subsequently in the Netherlands Aided Projects (NAP), Pani Panchayats (Pani means water and Panchayat means Committee i.e. a village level water committee) were formed. So far such committees have been formed in 77 villages. Pani Panchayat is a non-governmental/non-political body of six selected members from different socio-economic groups. The structure of Pani Panchayat consists of two male members, two female members, the Sarpanch (village head) or Deputy Sarpanch and Lineman/women. The Pani Panchayats work as a watch dog to ensure the smooth functioning of services.

Female members from various groups are given proper representation in the committee. With a view to providing necessary training and awareness in the programmes, the State Government has established a State level Training Institute with the financial support from the World Bank (INR 30 million). The Institute was commissioned in 1988 which has by now gained national reputation. This Institute is catering to the needs of government staff, NGO functionaries and the beneficiaries. The members of Pani Panchayat and village women groups are also provided with training. Women staff in schools, in health and paramedical services and social service groups are also provided with awareness and training.

Priority is accorded in providing training to women participants/groups. Similarly departmental women candidates are provided with adequate training within country and abroad.

Income generating activities for women in NAP areas have been started with the support of RNE and an NGO (SEWA). The local women's craft activities are encouraged through providing them with basic facilities and market. The income generating activities include dairy cooperative, handicraft, salt production, charcoal production, nursery, afforstation, etc.

In addition to this, a large number of NGOs are working in various activities related to the Water Supply and Sanitation programme; the majority of them are concerning awareness and training with special emphasis on women.

To accelerate the activities of awareness and training in NAP areas, a separate cell known as the Socio-economic Unit (SEU) has been created with the support of the Royal Netherlands Government. Similarly, under the support of UNICEF funding, an IEC Unit (Information, Education and Communication) has been recently created. These units are working for social mobilization in the sector programmes. To monitor these activities a State level *Steering Committee* headed by Secretary of the department and district level *Advisory Committee* headed by Collector of the District are formed.

Under the Rajiv Gandhi National Technology Mission Programme of Government of India, a separate *Literacy Mission* for promoting literacy, through the Adult Education Programme, is started which is doing commendable work in rural areas of the State. This mission is concentrating more on education of female members. The State Government has made *girls' education free* at all levels. The Government of India and State Government, both, have also resolved recently to *reserve 33 per cent of seats for women* in all government elections and office bearers.

Once the water supply system is developed at village level, the operation and maintenance is required to be handed over to the respective village panchayat (administration). It has been experienced that due to various reasons, such O&M is not done properly by the village body and the systems fails, thereby defeating the objectives of the project.

The salient reasons of this failure are non-payment of water tax and lack of O & M skills. To overcome these difficulties, Pani Panchayats in a large number of villages under NAP areas are established. As mentioned above, women are also members of these pani panchayats. It is now envisaged to operationalize more and more Pani Panchayats for handing over the village level O&M to these Panchayats after training and equipping them for local maintenance and management. Special attention is being given to involve also women in this training process. An experimental project has been taken up in five villages and the members of the panchayat are provided with orientation and training. It is encouraging to note that women participants are taking keen interest in such activities.

References

o 'Women, water and sanitation' by Siri Melchior-Tellier, Prowess, UNDP, USA, Special issue of *Water International* Vol. 6 Sept. 1991.

o 'Community participation and women's involvement in water supply and sanitation projects'. (Occasional paper series). Published by International water and sanitation centre. The Hague, The Netherlands - August 1988.

o 'Role of women in water supply' by Ms. Anuradha S Gadkari of National Env. Engg. Research Institute (NEERI), Nagpur, India. Article published in *Journal of Indian Water Works Association* - Vol. 18 of 1986.

o 'The Emerging Power' A publication on participation of women in the Indo-Dutch Drinking Water Programme. Published by Royal Netherlands Embassy, New Delhi.

o 'Woman, Water, Sanitation'. Annual Abstract Journal No.3 Nov. 1993 published by IRC and Prowess.

o *Women, Water and sanitation* - A WHO Publication.

Developing participatory hygiene education materials

Mrs Jemima Dennis-Antwi, Ministry of Health, Ghana

HYGIENE EDUCATION has increasingly become an important and inseparable component of any water and sanitation programme especially in developing countries.

Hygiene education seeks to encourage a target audience to utilize the resources available to them to achieve optimum health. It also serves to empower people to initiate programmes or activities that will ultimately improve their health. The use of visual aids promotes the impact of this process. Visual aids are very effective tools in enhancing effectiveness of hygiene education. They assist educators in carrying out their tasks and also give them confidence and credibility to perform those tasks.

Developing visual aids involves a series of steps necessary to ensure that the products meet the needs of the target audience in relation to:

- Pictorial accuracy
- Personal relevance
- Comprehension of messages
- Appropriateness and understanding methodologies

The Kumasi Health Education Project (KHEP) has over a project period of three years proved that affordable community hygiene education materials could effectively be produced and used for education in the community and schools using health education agents who have been trained in the appropriate use of the materials for effective education. The experience of the Project in material development has broad implications for water and sanitation projects in Ghana and elsewhere.

Water and sanitation related diseases such as various types of diarrhoea, worm infestations, skin and eye infections and vector borne diseases account for most of the morbidity and mortality in developing countries.

Water supply and sanitation programmes generally aim to improve public health through the reduction of diseases that are water and sanitation related.

Until recently the main emphasis was on the provision of new and improved facilities with little emphasis on hygiene education, the absence of which leads to poor links between facilities and practices with regard to the use, care and maintenance of facilities, use of safe water in sufficient quantities and the safe disposal of waste water, human and other solid waste. Hygiene education is most often directed at specific target groups. The use of visual aids enhances their attraction to hygiene education activities and to remember the messages communicated to them. An evaluation of visual aids showed that a combination of words and visuals was remembered some six times better than words alone and three times better than pictures alone (MacDonald et al, 1984).

Hygiene education programmes everywhere have made extensive use of posters and flipcharts. As well, a wide range of audio-visuals are used including models, printed illustrations and photographs, flyers, newspaper articles, radio talks etc. Other means such as drama, songs and games are also being used, as well as real objects such as demonstration latrines and water filters (Burgers et al, 1988).

Sometimes both types of media can be integrated successfully, but this requires careful assessment of the available media and their compatibility (Ling, 1986).

Often, people acquire knowledge and understanding more easily by showing and handling real objects. Therefore, hygiene education programmes may benefit from using the real thing or if not, then models.

The experience of KHEP in developing participatory hygiene education materials

Developing hygiene education materials was a very vital component of the project's activities and it involved going through a series of stages in ensuring that the end products meet the need of the target audience.

The stages include:
(1a)   Identification of methodologies
(1b)   Ideas generation
(2)   Material design
(3)   Review of prototypes
(4)   Pretesting
(5)   Production

Preparatory phase - identification of methodologies

To ensure that appropriate methods are selected in developing hygiene education materials, a focus group discussion is held prior to generating ideas for the material in question. The discussion usually involves facilitators, graphic designers and selected health education agents identified within the metropolis. These mostly include: teachers, public/community health nurses and environmental health officers.

Issues discussed include:

(a) Logistical problems anticipated in implementing a health education activity.

(b) Material requirements eg. size, colour durability and methodology preferred.

(c) The target audience that the material can be appropriately used for.

The information derived is used by the facilitators to pre-select appropriate methodologies for developing the material, when an agreement has been reached on the above.

Ideas generation

The objective of this stage is to provide ideas and information for the development of an hygiene education material eg. a flipchart on diarrhoea.

To achieve this objective, a one day workshop is held comprising a maximum of twelve participants including two facilitators and two graphic designers. These participants will be asked to divide themselves into two equal groups with a facilitator and a graphic designer in each group. Their role is to guide the participants in providing adequate clear and detailed information to facilitate the designing of the materials. After the exercise, a plenary session is held whereby the two groups present their contributions for a consensus to be reached.

Design of materials

At the end of the ideas generation workshop, the facilitators and graphic designers collaborate with each other to refine and provide a pictorial representation of the ideas and information provided by the participants. This is spread over a period of two weeks. It may be necessary here to employ an additional artist if the work is too much for one artist.

Review of materials

To allow for further development of the prototype material developed in stage two, a second workshop is held comprising the same people who attended the first workshop. They are divided into the same groupings as in the first workshop with a task to examine firstly the prototypes presented and based on the information they gave in workshop one, and to make the necessary changes with the help of the designer. In a plenary, the group will again present their recommendations and justify them.

A member of each group will then be asked to carry out a mock health education workshop to determine the suitability of the material. There is a final period of material modification and finishing of pictorial content following the second workshop by the designers prior to pretesting.

Pretesting

The purpose of pretesting is to ascertain the pictorial accuracy, personal relevance, comprehension of mes-
sages, appropriateness and understanding of the methodology. For established methodolgies, the pretesting is mainly concerned with the visual accuracy and message content (Laverack, 1993).

Pretesting is carried out in three steps.

The first step involves showing the final prototype of the material in pencil to a group of the target audience in the community for their recommendations. Based on their recommendations, the prototype is further reviewed by the facilitator and designers.

The second step involves presenting the reviewed prototype in a black outline which will include more detail and clarity to another group of the target audience for their reactions. These are considered for further review of the prototypes.

The final step is when colour is added to the prototypes and taken through the same procedure as in step one or two.

Production

This is carried out when there is the satisfaction by all interested parties that the contents and messages are being accurately presented by the materials. It involves separating out the colours that will be used for printing the materials before a press is contracted to do the printing work. The whole process of material development takes not less than three full months. Through the procedures outlined, the Project has developed a range of participatory hygiene education materials that are used for education activities within the metropolis and other non-governmental organizations. Presently efforts are being made to provide these materials to be used on a national scale in all health facilities.

Some experiences worth sharing

(a) Organizing the workshop

Participants could conveniently be selected from governmental organizations present in the locality, for example Health Ministry, Education Ministry, Department of Community Development, depending on the target audience and the subject matter.

(b) Pretesting

- Organizing the community is the most time consuming part of the pretesting exercise.
- Difficulty in controlling large numbers of people who congregate during the exercise.
- Dominance by particular participants.

(c) Reporting

Reports should include enough details to allow a focus group to evaluate the effectiveness of the materials for the target group.

(d) Material production

The following experiences were found to reduce cost and improve durability and quality of the graphic materials.
- Using colour tones at the colour separation stages.
- Using metal or plastic spines to bind materials.
- Providing gloss finish to the front cover of materials.
- The use of colour permanent and washable cloth for colour posters.
- Collect printing plates and store correctly for future production. (Laverack, 1993)

Some participatory hygiene education materials produced by the project:

Flash cards
Mosquito Control; Prevention of Diarrhoea; Prevention of Roundworms; Waste Management; Food hygiene; Personal Hygiene and Dental Hygiene.

Flip charts
Prevention of Diarrhoea and the Worm Calendar.

3-Pile sorting cards
Water Supply and Diarrhoea.

References

1. Burgers, L; Boot, M; Van Wijk-Sijbesma C (1988) 'Hygiene Education in Water Supply and Sanitation programme': Literature review with selected and annotated bibliography (Technical paper series; No 27). The Hague, Netherlands International Water and Sanitation Centre.

2. Laverack, G. R. (1983). *The Kumasi Health Education Project: Health Experiences in the Kumasi District, Ghana,* ODA.

3. Ling, J.C.S. (1986) *Health and the Media.* World Health March 1986, 18-19.

4. MacDonald, I and Hearle, D. (1984) *Communication Skills for Rural Development.* London, UK, Evans Brothers Limited.

Water and sanitation — a gender approach

Vijita Fernando, NGO Water and Sanitation Decade Service, Sri Lanka

IN THE BEGINNING water and sanitation (WS) projects tended to focus almost exclusively on physical works. The people for whom they were intended were mere users or beneficiaries. This thinking changed fast, fortunately, and at least a decade before the UN decade for water and sanitation was declared in early 1981, community participation began to be identified as the key to the success of WS projects. But community participation became a male affair and women had no active role. They became 'users' or 'target groups' for health education with only the community men being involved as leaders, committee members and caretakers.

For greater efficiency and effectiveness of WS projects it became quite clear in the eighties that women needed to play a greater role. Their views had to be listened to and their participation was vital if WS projects were to bring extra benefits for the community and women. The realization of the value of women's participation created a demand among projects for practical guidelines and how they could be brought effectively into planning, implementation and maintenance processes. These guidelines proved very effective and worldwide, many projects and programmes had glowing stories of the effectiveness of projects with active and full women's participation.

Where were the men now? There was a tendency to ignore the roles of men and this meant that focusing more on women's roles reduced the responsibilities of the men. Extra focus on women also overburdened women who already had the hands full, especially in Third World situations. It also brought cultural problems. The need for women and men to share the decision making, the work and their functions more equally was becoming apparent, resulting in what we now call a 'gender approach'.

It is rather difficult to get the concept crystal clear. In simple terms, gender, as against sex which is a biological difference, is the result of a socialization process which assigns to women and men certain aptitudes, roles and responsibilities leading to certain forms of behaviour. Gender is the social and changeable difference between man and a women in a particular social situation.

In the gender approach we assume that the community — men and women — are the agents of their own development, with development agents in a supportive role. In situations where women are in a subordinate group women can easily be denied an active role in development with the self determination of the community becoming the self determination of the men. What is meant by gender approach is the aptitudes, roles and responsibilities of both men and women are taken into account requiring an open mind and aiming at the fullest participation of both women and men. Here we look at the two concepts — women's involvement and the gender approach. Women's involvement means that due to their disadvantaged position there is a perceived need to uplift them and bring them into the mainstream of development. Programmes focusing on women's involvement often have a negative effect as they tend to increase women's workload.

The gender approach, on the other hand, envisages a situation with more equality and justice, also taking into account existing aptitudes, roles and responsibilities of women and men. This approach gives more opportunities for women, make them share their burdens and recognize women as equal partners.

It is a working together for water and sanitation in its simplest form!

'Together for water and sanitation' is a project I would like to talk about, perhaps one of the very few efforts to introduce a gender approach to WS that has taken place in Sri Lanka.

In September last year (1993) 17 participants from seven Asian countries and Yemen worked together for two weeks in a hotel on the outskirts of Colombo close to the beautiful beach to develop perhaps the first manual giving guidelines for a gender approach in WS. This project was by the International Reference Centre of the Netherlands and co-ordinated by the Sri Lankan NGO consortium, water and sanitation decade service.

We were 16 women and one man. They came from Bangladesh (the man!), India, Nepal, Sri Lanka, Yemen, Bhutan, Philippines, Pakistan. The Hague and Sri Lanka provided the facilitator and co-ordinator.

I think before I go on to tell you something of the process of producing the manual, I should answer the question I myself asked when the process was launched. Why a manual? The rationale was that though individual countries in the region — and elsewhere in the world — were trying out new and innovative methods in gender and other aspects of WS, other countries were not aware of events taking place even in the region. The manual would document such experiences providing a source of information from other contexts and other experiences on how they were trying to introduce a gender approach to WS.

Also, despite the growing concern and urgency of an effective gender approach in WS, no practical field tested guide was available which had a strong base of experience from the field.

Visualizing and producing such a guide as the one we did was a 100 per cent participatory effort. It would include participants' methods of working on gender concepts at field level, organizational methods of sensitizing staff on gender issues and at policy level, how implementing agency fieldworkers influence policymakers. It is through the eyes of the participants that the manual would view 'women's involvement/gender approach' first broadly and in a wider context and more specifically as related to WS.

As we worked through the day and sometimes half the night, we were a closely knit group, sharing information on seven countries on a variety of projects and a hundred ideas on how to produce a manual!

With the background of the participants' work in WS clear, we identified problems related to women's involvement each encountered in his/her work. Problems were clarified on a problem tree which helped in the identification of the root of the problem hammering a gender approach. It was easy to see the nature of the problem — these were not always related to lack of water. Quality of water, sanitation and the environment, drought which increased women's workload and increased women's workload and health problems caused by constant carrying of water on the head — all these stood in the way of women's true participation.

One hard fact emerged — it was impossible to make general statements about the role of women in WS. The situation in each country differed even when the countries were in the same zone.

However, it was possible to come to some conclusions. The ultimate aim of women's involvement is to achieve a more equitable society — in relation to the burden of women's work burden, decision making and planning, access to paid official positions. WS projects address practical needs of women. But through the working methods used, strategic needs can be addressed. Women's involvement thus is not a mere contribution of labour. It is access to resources, decision making and management tasks. It should not be a process to overburden women further and reinforce their gendered roles. To achieve this men need women to be sensitized on gender issues, the roles of women and their own roles.

Shared experiences showed us that while the position of women varied within the region, the lot of the rural woman is an unenviable one , with their triple burden of home, work and community. They are overtired, undernourished with too much work and too many pregnancies. Often they are a prey to strong forces within the family and the larger society outside the home. The woman has little voice in matters that concern her or her children outside the home. She hardly ever makes decisions in public affairs that concern her vitally, as WS, the environment, her health. Such decisions are left to the men. She has no voice even in the physical well-being of her body, often her only possession. Tradition and culture shackle her. Illiteracy isolates her. Girl children are often unwelcome as they entail a tremendous financial burden on the family.

Scarce water places special burdens on women who do not always realize that they themselves contribute to their overburdening as in cases of deforestation — resulting in less water and longer distances to carry.

All of us agreed that empowerment of women should be a common goal for women to take charge of their own lives, be less dependent on men to have their voices heard. We spoke of the importance of employment to increase their decision making capacity to influence changes that reduce their drudgery. Hopeful silver linings emerged.

With these facts at the backs of our minds, we prepared the guidelines for the manual — identifying tools, case studies, checklists for policy development, gender sensitizing programmes for project staff and guidelines for the integration of a gender sensitive approach in all project phases. The manual also would include a gender specific baseline data formulation, environment assessment, planning with the community, developing gender sensitive materials, task decisions which would increase gender awareness at community level, increasing women's self-esteem through economic activities, selecting the right technology, training for operation and maintenance, spare part supply by women's groups, a women's water book and some real soul searching in the section 'Do women really participate?'

The manual also has a sprinkling of case studies from the various participating countries, list of books and videos which might be of help, and a note on funding agencies.

As I write this paper the manual is still in draft form. Those who got it together have gone through the draft and provided comments, criticisms, deletions, additions to perfect a labour of love, painstaking produced by this group and ably put together by IRC's Eveline Bolt.

Hopefully by the time I present this paper the final version may be available for us to look at.

I certainly hope so!

Empowering women to manage watsan technologies

Ms. Ruthy C.D. Libatique, Kabalikat ng Pamilyang Pilipino Foundation, Inc., Philippines

LACK OF WATER IS A PROBLEM. But women as the principal users of water by virtue of their traditional role in the household are more affected than the men. Some even perceive the problem as more theirs than men's. Men who are out on farmwork or other labour lack the time to install a water supply system in their homes or at least in the neighbourhood. This is the situation in which, the women of Capiz found themselves in 1988.

Capiz is a province in Central Philippines where there are abundant rains during the wet season. Capiz Development Foundation (CDFI), a local non-government organization, identified the Ferrocement Rainwater Catchment Cistern Technology (FRCS) as appropriate for the province. Since 1984, over 300 Ferrocement Rainwater Tanks (FRT) have been constructed in Capiz with assistance from the International Development Research Centre (IDRC), United Nation's Children's Fund (UNICEF) and the *Tulungan sa Tubigan Foundation (TSTF)*, another Philippine NGO. Community participation is a major approach CDFI and TSTF are applying in the construction of these facilities. However, constructions are often delayed because the menfolk could only attend to the job during their off-days from farmwork or employment. The women were left to wait until 'dream tanks' were built. Because the problem had long existed and fed up with waiting till their men could put up the facility, the women resolved that they themselves would build their waste tanks. The women asked to be taught how to build the tanks. Why not? That was not a bad idea at all. But how do you teach the women, who could barely read and write, technical knowledge and skills on construction?

Training the women was the first option thought of by the implementers. But training women for every construction job would prove to be very expensive and time-consuming. After several brainstorming sessions the implementers thought of developing a manual that will teach people, men or women, how to construct and maintain a ferrocement tank in a nontechnical manner. The idea was offered to the IDRC for consideration. IDRC liked the idea at once, and called on *Kabalikat ng Pamilyang Pilipino Foundation Inc.*, a non-government health organization known for material development and health research to develop the manual on the installation, operation and maintenance of ferrocement tanks for CDFI, with assistance from TSTF.

Kabalikat immediately put together the team for the manual development. A materials development specialist (the author of this paper), an artist and a project assistant composed the Kabalikat team. A woman engineer from the provincial government of Capiz technical staff from TSTF and community development workers from CDFI helped the team in the various steps of development.

The first problem I encountered in the development of the manual was how to write the manual when I myself had zero engineering knowledge. Secondly, if I had the engineering knowledge, how would I convey this knowledge in a way the layman would understand and find interesting? How do I teach people especially women how to construct a ferrocement tank? One thing was certain, the manual has to be 'women friendly'.

The team drew up the plan of action. First, we had to learn how to construct the tank. So the artist and I went off to Capiz for a week to experience how to construct the tank from its initial to the final stages. We measured, mixed mortar, tamped and plastered the walls, along with the menfolk of the community. The first day was a nightmare. I could not understand what my workmates were doing. I was also hesitant to ask the menfolk every now and then for fear of holding up or slowing down the work. Fortunately, the lady-engineer patiently explained

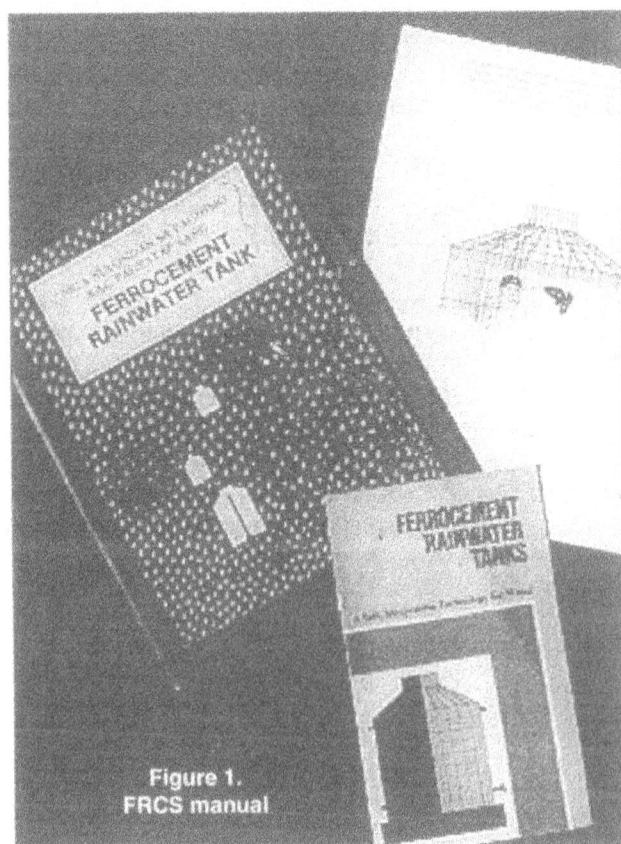

**Figure 1.
FRCS manual**

to me the steps of construction and the reasons for them. By the third day the artist and I felt very much part of the group. By the sixth day, our ferrocement tank was complete. In between participating in the construction, I had to stop and write key steps that would be helpful in the writing of the manual. The artist meanwhile made sketches.

After practical exposure to the building of the tank, I wrote the manual. Every now and then the step by step procedures were referred to the lady-engineer to check their technical accuracy. After outlining the steps I sat down with the artist to conceptualize the visuals of the manual.

The first draft was subjected to technical review by the lady engineer and the TSTF staff.

The visuals were drawn, but no instructions were put in the first draft to see whether the target audience would understand the visuals even without the corresponding text. Groups of men and women were recruited for the first pretest to determine comprehensibility, appropriateness and clarity of the 'visuals only draft'. The results were analysed and corresponding revisions were made.

The second, third and fourth pretests were on the 'visuals with test drafts' with progressive improvements after each pretest. All the pretests used the focus group discussion approach. The second, and third pretests were conducted among men and women groups while the fourth pretest was with a *purely women group*. All the previous pretests paid special attention to the comments of the women. All the comments were responded to and validated in the subsequent pretests.

After every pretest, the artist and I went back to the drawing board to make the material more comprehensible and appropriate for lay persons especially the women. After four pretests, the manual was ready for field testing. The field testing required women to use the manual in building their own watertank. CDFI provided all the materials needed for the tank. There were 10 women and two men participants in the field test. I explained to the fieldtest participants that it was the manual itself that was being tested and not them. The women participants studied the manual, talked among themselves and proceeded to construct the tank. While building the tank, one woman who appeared to be the leader of the group was holding the prototype manual and reading it to the group. Every

now and then the group would gather around her to read the instructions.

While the fieldtest was going on the lady engineer was observing how the construction was proceeding while I was intently observing which portion of the manual the participants appeared to be having difficulty in following.

The digging for the foundation was done by the two male participants but the women participated in all other aspects of the construction.

Before the day's work is over, the author and the lady engineer discussed with the fieldtest participants what transpired during the day and what portions of the manual they found difficult to follow. The comments were duly noted for subsequent improvements later.

On the seventh day a brand new 10 000 litre tank was standing in the neighbourhood.

The 10 women and two men were proudly eyeing their handiwork built by following the construction manual. The participants could hardly believe they did it without supervision from a construction foreman or a mason. They were in fact asking if the tank that they built was strong. A final discussion with them was conducted to summarize and validate findings of the previous days.

Finally, the lady engineer was introduced to them as the engineer who was silently supervising their work. When they learned that an engineer was actually approving their day to day output, they started congratulating themselves. The women were very proud, at the same time astounded that they could build a tank. Now they are saying, give us a manual on how to construct a house and we will.

Today ferrocement tanks of various sizes dot the Capiz landscape. The women of Capiz are specially proud because they built their own tanks. The final version of the manual were distributed and used by the women in building their water tanks.

The lesson of this exercise is that audience-oriented materials can be more effective instruments of empowerment.

Material developers or writers should involve the target audience. One must not forget that he/she is writing the material for them and therefore consultation with them is vital. Doing so empowers your audience and soon the material shall empower them.

Community infrastructure programme in Pakistan

Jelle van Gijn and Brian Ellis, O'Sullivan & Graham Limited, Islamabad, Pakistan

IN TEHKAL BALA, *an urban community within Peshawar in Pakistan's North West Frontier Province, a group of local people is working hard to surface their own streets with brick and stone, after waiting for many years for the local council to do so. The streets and footpaths have good lined drains, and the people no longer suffer from the unhealthy environment created by the muddy streets and the blocked drains. Funds for these works have come from the savings of the people from Tehkal Bala, with support from local and provincial government.*

Further north in the Swat valley, the residents of the rural settlement of Ghalegay are constructing diversion weirs and check dams in stone masonry and gabions, to control flash floods coming down from the hills above their village. They have surfaced streets and footpaths, re-built local drainage system and relocated water supply mains.

The quality of the workmanship is remarkably high, with the help of local artisans. The costs have turned out to be about a quarter less than the cost estimates based on government rates, a cash benefit returning directly to the community.

Introduction

Pakistan has made impressive progress during the past decade in the development of its industry, commerce and major infrastructure. A national highway network is under construction, telecommunications have improved dramatically, industry is expanding away from the traditional sectors. Against the background of these remarkable achievements stands in stark contrast the relative neglect of the social sectors, in particular in the rural areas. Illiteracy among women in certain rural districts is as high as 96 per cent. Only a fraction of urban and rural population has access to adequate sanitation. Primary health care, education for all, shelter and basic infrastructure have now become the focus of government programmes, with support from international agencies.

A national programme of pre-investment studies in the Shelter sector pointed to the deep-rooted problems that contribute to the current weaknesses in the provision of housing and infrastructure:

- inappropriate public finance mechanisms for basic infrastructure

- lack of an active housing finance industry

- inappropriate technical standards and quality control.

The programme tried to develop new ways to deliver basic services to low-income communities in both urban and rural areas. The programme encourages the population to take an active part in improving their own physical and social environment.

Pilot projects were designed to try out new approaches to improve housing and infrastructure on a small scale in different provinces. This paper describes the successful experience with one of such programmes, in which community organizations are actively and directly involved in planning, designing and financing basic infrastructure: the Community Infrastructure Programme (CIP) in Pakistan's North West Frontier Province (NWFP). CIP seeks to improve basic infrastructure in many urban and rural settlements throughout the Province, financed by the Government of Pakistan, the World Bank and Switzerland. Guiding principle of CIP is that beneficiaries contribute to investment for local infrastructure, and take full responsibility for the cost of its operation and maintenance.

The process has introduced fundamentally new ideas and roles to all actors in the programme of development: to local and provincial government, to international aid agencies and to the local communities themselves. A project preparation cycle for infrastructure improvements — traditionally involving the urban planner, the engineer and the financial analyst — now includes the community development planner and the social organizer: not merely to seek community participation and improve awareness, but as an integral part of the planning process.

The organization

Pilot projects are prepared by a multi-disciplinary team working within a Project Management Unit (PMU) established as part of NWFP's Department of Physical Planning and Housing, in collaboration with the Department of Local Government, Elections and Rural Development. The community development group consists of male and female social organizers (SOs) specifically recruited and trained for this programme, headed by a community development planner. SOs have different educational backgrounds, and have degrees or diplomas in social sciences as well as in languages, law or geography. They have been trained in the skills of social organization, with lessons drawn from other community participation programmes in NWFP and elsewhere in Pakistan. They have been selected for their ability to listen to people and to encourage them to organize their community to achieve improvements in their living environment. Within the cultural climate of NWFP, female SOs often need permission from their family to undertake a job such as this that requires them to travel independently.

The community development mechanism

The central theme behind the process has become that any improvements to local infrastructure should be planned and built together with the people who will use these works. The programme needs to find out what the people want most, what they can afford and how they want to achieve it. To find a common opinion among a group of people, a representative community organization of some type needs to be found or formed. This is where the process of project preparation starts: community mobilization.

Community mobilization

This aims to establish a group within the community which can act as a representative body for voicing opinions and acting for the community: a Community Based Organization (CBO). Social organizers — both male and female — begin to visit selected communities and make contacts with individuals and representatives of local social and cultural groups. They need to understand the structure and distribution of power and influence within a community. It is the role of the SOs to ensure that any established elite does not feel threatened by the initiative, but that they understand the aims of the social organization, and that their influence is mobilized to support the process. Certain communities already possess an active CBO. SOs will introduce the infrastructure programme to the CBO and assess their interest in such a programme.

Socioeconomic surveys

Once a provisional interest by the community in the programme has been established, the team conducts a socioeconomic survey. The survey serves many purposes. It gives the people an opportunity to state their priorities and concerns, on health and education, on services and housing. Simultaneously, it improves the SO's understanding of the community and allows a chance for discussions with individuals and leaders. The surveys provide the basis for later planning and design, with data on infrastructure needs and priorities, willingness to pay for additional services and affordability. At a later stage, engineers will make a preliminary visit, to gain a first impression on the scale of infrastructure needs, the availability of trunk infrastructure and the level of investment required to improve services.

Community based organization

The objective of social organization is to create strong representative groups at the community level. The role of the CBOs — once established — is broader than only for infrastructure improvement: the aim is to create a self perpetuating institution through which the community members can work together to manage their human and material resources to reach higher standards of living. It is only through such maturity that the communities can maintain the CIP package delivered.

The creation of a CBO needs to be formalized to link with government activities, as the CBO will eventually manage collective funds, and award or execute contracts. Name, constitution and procedures, the election or selection of its general membership, executive body and office bearers need to be defined and registered.

Through internal consultation the CBO defines the common needs, and agrees on priorities. For each priority, the initial willingness on cost sharing is agreed between the CBO and the project. Guiding this process, the social organizer must ensure that the needs identified by the community match the targets of the CIP. Inevitably this has lead to some disappointment and misunderstandings as some communities have expressed a need for a girls' secondary school or a district health centre. Although the current CIP cannot provide such facilities, the programme has now established linkages with UNICEF and other initiatives that could respond to these needs. The CBO is thereby providing a basis for a wider role and to become the focus of other developmental activities, such as health and hygiene awareness programmes, and support to women and children.

Preliminary designs and cost

The list of priority infrastructure need improvements as defined by the CBO forms the basis for the preliminary engineering design work. A topographic survey and a detailed survey of existing infrastructure result in 1:500 maps used by engineers for an initial design of improvements in all selected sectors. Designs are based on sets of *design standards*, adopted as appropriate for the nature of this programme [see box 1].

CIP focuses on local or *internal* infrastructure: provisions that are within the boundaries of the target settlement, and that can be improved without the need for major trunk or *external* infrastructure. Some external infrastructure is included where it is considered essential in supporting internal infrastructure and in achieving the global programme objectives of improving living conditions for low-income settlements. The costs for any such trunk infrastructure will be fully financed by government. Special funding arrangements have been introduced for sanitation improvements because of the still developing awareness on the need for improved sanitation facilities (see box 2).

Project cost estimates are based on unit rates accepted by government. From preliminary project cost estimates, the financial analyst prepares a first financing plan, according to global cost sharing principles accepted for the programme (see below). An assessment of household affordability — by which households will not be expected to spend more than 3-5 per cent of their combined income on the programme — decides the scope and phasing of

Box 1. Design standards and target service levels

Design standards for CIP are based on the needs and environment of low-income communities. An additional consideration is that works need to be built by communities themselves or by small local contractors. This has implications for the choice of material and for standardization of the range of designs. The standards are considered to be incremental and in future could be upgraded.

water supply:	house connections or standposts within 100 m of every house. Distribution system in GI pipe, dia. 25 to 100 mm. Per capita consumption through intermittent supply of 60 lcd (rural) to 75 lcd (urban).
drainage:	(i) concrete trapezoidal channels cast in situ, or (ii) rectangular brick channels with cement sand rendering, on concrete base.
sanitation:	**on plot:** double pit pour-flush latrines; demonstration project through provision of slab and twin-pit as incentive. **settlement ponds:** pre-treatment of heavily polluted storm water before discharge.
solid waste:	local concrete or brick work containers within 80-100 m of every household. In urban areas, collection for final disposal through city-wide system. In rural areas: local disposal through burning or burial, with extensive reuse.
roads & access:	**access roads:** flexible pavement asphalt premix or double surface dressing; **minor streets:** brick-on-edge or (in hilly areas) concrete; **footpaths:** flat-laid brick on cement sand, or concrete.

Box 2. Sanitation

The sanitation component is designed to provide a budget allowance in the loan for the implementation of a programme of introducing improved twin-pit pour-flush latrines in all communities. The component has specific significance and urgency in these communities, as open storm water drains are inevitably contaminated with sullage as well as faecal matter. Adequate treatment of this waste water cannot be realistically expected on a large scale. Upgrading will therefore need to focus on a reduction of the faecal contamination at source. The demonstration latrine improvement component aims to achieve this, through a combination of improving health awareness, training in construction techniques, demonstrating the available options, and providing an incentive through a 25 per cent cash subsidy on the cost of constructing slab and twin-pit.

The project cost estimates are based on providing on average an additional 30 per cent of the households with a latrine by the end of the project, in the year 2000. Costs are based on a construction cost of Rs 4000, which excludes the cost of the superstructure. 75 per cent of these costs will be paid directly by the individual household. This contribution is *not* included in the affordability calculation for assessing the community contribution to the remaining infrastructure improvement programme. The 25 per cent subsidy will be financed by IDA, through the provincial government.

the programme. The capital cost per household is about the same as that of a new TV set. The outcome of this first round of design and cost estimates is the monthly cash contribution that will be required from all (beneficiary) households in the community. This information — preliminary design and corresponding household contribution — provides the basis for a key meeting with the community: the presentation of the plan.

Presentation

The presentation meeting provides a broad forum of discussion. Usually chaired by the Chairman of the CBO, the proposed plan is explained to the meeting. Different options are discussed, with the implications for required contributions by the households, for both initial investment and the recurring obligations for operation and maintenance. The role of the SO and the engineer is to introduce technical issues in an understandable way to an untrained audience, who may help to decide the size, scope and phasing of the programme.

Following the discussions, designs and cost estimates are revised. If major changes follow from the discussions, a second presentation may be held. The plans as adopted proceed to detailed design and the preparation of tender documentation.

Detailed engineering design and contract documentation

Contract documentation for implementation of the CIP follows procedures laid down by the Government of Pakistan. At the same time, however, procedures have to support the principles of CIP, i.e. that contracts can be issued to small local contractors or to community groups. In addition, procurement guidelines of the IDA need to be observed. Developing procedures that can satisfy all these requirements proved one of the main challenges facing implementation of CIP.

Civil works contracts included in government annual development budgets in Pakistan require detailed engineering design documents and cost estimates, to be submitted in the form of a PC-1. The cost basis is a very detailed bill of quantities with unit rates derived from base materials and unit labour rates. Variations of either quantities or costs beyond a narrow band require resubmission of the PC-1 and can lead to serious delays in the processing of payments. The experimental nature of CIP — involving communities in the planning and building process — inevitably leads to changes in costs and quantities. Unless government procedures could be interpreted with more flexibility, interruptions in the flow of funds would seem inevitable, with the risk of alienating communities from the programme.

For the initial batch of twenty CIP sites, separate formal PC-1s were prepared, with full engineering drawings. On the basis of approval of these individual PC-1s, an *umbrella PC-1* was formulated, representing the over-all cost of the first phase of the programme. For the next phases, a more global approach can be adopted based on experience of the first batch of sites. An element of flexibility is required to support the true characteristics of the CIP. Eventually, an overall budget allocation to a certain community will be based on a preliminary design and accepted unit rates. This will allow the community to adjust to the inevitable changes that will arise in the quantity and type of works, as well as in their costs. The community can do such adjustment by changing contracting arrangements, modifying their priorities or modifying the designs. It is essential within the principles of the CIP that this flexibility is built in and maintained.

Financing: who pays what

The principle of cost sharing is fundamental to the concept of CIP. Contributions by beneficiaries to local improvements, however small, ensures that the investment in infrastructure will be maintained. Participation in the design and planning provides an incentive to the population to mobilize their resources, and commit savings to improvement of their own living environment.

Present funding arrangements are based on a 20 percent contribution by the community for all internal infrastructure. Local councils pay for another 10 per cent, with the remainder funded by provincial government, partially from the IDA Shelter Project loan. Trunk infrastructure is funded completely through the Provincial Government. Within a parallel ongoing programme of strengthening the resource base of lower levels of government, it is the intention that the share of local councils in funding CIP will gradually increase.

CBOs are responsible for collecting funds from within the community, setting levels of monthly cash contributions required from households. Internal arrangements can vary, and often those with higher income are found to contribute more. A community has to demonstrate its commitment to the programme by collecting their share of the first batch of the works before building can start.

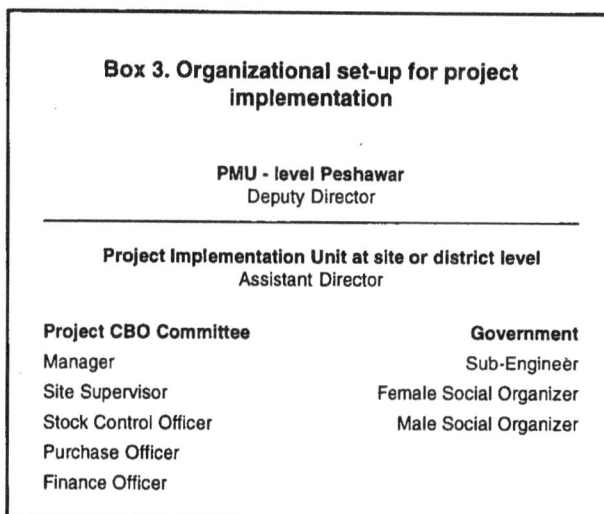

Box 3. Organizational set-up for project implementation

PMU - level Peshawar
Deputy Director

Project Implementation Unit at site or district level
Assistant Director

Project CBO Committee	**Government**
Manager	Sub-Engineer
Site Supervisor	Female Social Organizer
Stock Control Officer	Male Social Organizer
Purchase Officer	
Finance Officer	

Building the works: contracting and supervision

CIP intends to create an opportunity for small local contractors or individual craftsmen within the community to be involved in the building process. Contractors are encouraged or obliged to employ local people as labour wherever possible. Apart from the generation of local employment, these procedures provide a greater control on expenditures and quality, as those involved in building will benefit directly from the completed works.

Implementation arrangements have been set up to respond to the requirements of community control on expenditure, whilst still operating within the procurement guidelines of government and IDA. Works have been divided into three categories: (i) Type A contracts for primary infrastructure should be awarded to large contractors (registered as Class A or B), (ii) Type B contracts for annual packages of internal infrastructure can be awarded to Class C registered contractors, while (iii) Type C contracts for small and simple infrastructure can be procured through direct contracting by the community itself.

The flow of funds for financing of the project is a sensitive subject in a highly politicized society with some distrust on the part of the population in government management of public funds. A fundamental step is the creation of a *community project account*, with representatives of local government and the CBO as joint signatories. After a formal start to the project — with the signing of a Memorandum of Understanding and a Community Finance Agreement — funds can be transferred into the project account from both the community contributions and local and provincial government. The amount of money transferred corresponds to works planned for a given period, at project cost estimates based on agreed unit rates. In practice, community contracts will allow works to be built at lower cost than estimated from government unit rates. These savings accrue to the community, enabling building up of a fund for further local development work.

Responsibility for implementation planning and management, quality control and accounting is shared between a group of CBO's representatives and representatives from local and provincial government. This arrangement reflects the shared funding of the project.

Running the works: operation and maintenance

One of the basic principles of CIP is that the community should take full responsibility for operation and maintenance of the facilities provided through the programme, whether in cash or kind. The type of community involvement in this will vary between urban and rural settlements, and from sector to sector. The outline agreements suggest that the community provides free labour for routine cleaning of drains and waste collection points. Cash contributions are required to pay the electricity for tube-well pump operation, and build up a fund to cover equipment repair and replacement. The initial financing agreement already defines these obligations. As implementation has only recently started, the programme has little practical experience with how arrangements will materialise in practice.

Conclusions

So where are we now with CIP. After a long leading up period we are now seeing very encouraging signs at implementation. After months and years of breaking down old divisions of roles within the planning process, and expanding the understanding amongst government and lending agency bureaucrats on more open-ended approaches to project financing and definition, we are now seeing the enthusiasm and commitment of the local population at the receiving end of this preparation. There is still a long way to go. Traditional attitudes still guard the interests of government officials and the contracting industry. International procurement rules still tend to favour the strict definitions in project approach. But changes are now perceptible, at a small scale. Success has been demonstrated. Community involvement in planning and building ensures that what is built is really needed. Getting the community to pay part of the costs ensures that works will be looked after. Involving those who pay directly in building reduces wastage and unnecessary overhead. Allowing freedom and flexibility in design and choice of materials — within set quality limits — allows real savings to be made. Introducing community contracts encourages the use of local skills and the benefit of local experience. Expanding the role of community organizations to health and education gives access to the benefits of programmes of other agencies and NGOs, and advances the involvement of women in the development process. Involving communities in planning and implementation will eventually benefit local government.

The NWFP Community Infrastructure Programme — together with other encouraging new schemes in infrastructure and shelter development in Pakistan — provides a promising beginning to new approaches which can serve as inspiration to similar initiatives to improving living conditions for low income communities elsewhere.

References

[1] This paper reflects the views of the authors only. It may not necessarily reflect the position and views of Government of Pakistan, Provincial Government of NWFP, the World Bank, Swiss Development Cooperation and UNICEF.

SECTION 2

INSTITUTIONS AND MANAGEMENT

Institutional development, Brazil

Maria de Fatima Dimas Carteado and Richard Franceys, WEDC, UK

TO ENSURE THE safe provision of water and sanitation it is necessary to have effective and efficient institutions. This paper investigates the institutional changes in the urban water supply and sanitation sector in Latin American countries as different generations have sought to find the best approach.

The study focuses on Salvador in Brazil, covering the period 1853-1990. It is thought that with some differences due to local peculiarities, the pattern in other Latin America countries has followed a similar process.

Institutional changes in urban Watsan in Salvador, Brazil

The urban water and sanitation services in Brazil have been subjected to major institutional changes since the middle of the nineteenth century when the large cities first acquired a public water supply for the improvement of public health.

Private concessions

At the beginning, the service was run by private companies, mainly British or Portuguese. In Salvador, Bahia, the service was run by a Portuguese private company that had a concession from the council. Initially the company did not have a monopoly and the majority of the population continued using water sold by water vendors from fountains. The concessionaire therefore managed to persuade the Council to grant them a monopoly.

The concession contract was similar to the ones already being used in France whereby the private company invests in plant and equipment and a distribution network and charges the customers directly. The tariff was negotiated with the Council rather than being set directly by the company.

The contract with that first company lasted 50 years but was not renewed as the service by then was not considered satisfactory and the company and the Council could not agree about the setting of tariffs. The Council therefore bought the assets of the 'Companhia do Queimado' and by putting the service out to tender a contract was agreed and signed. The investment for upgrading the system was financed by the council through a loan from 'Banque L'Union Banque Parisienne'. There was a change therefore from what we would now term a 'Build, Operate and Own' (BOO) concession to a management contract where the concessionaire is not responsible for capital improvements.

Government department

Twenty years on, the poor quality and worsening service gave the regional government an excuse to take the concession from the Council. They decided to run the service as a decentralized government department. The staff of the water department were civil servants and the department had no financial autonomy, its budget being part of state budget. The Regional Government attached the water department to the Health Secretary with the intention for it to be self supporting through customer tariffs.

There was an emphasis on constructing standposts for the low-income householders and increasing the number of household connections to provide a reliable service for those who were paying.

Subsequently the water department was moved to the Building Secretary in the early 1940s, showing a change in perspective. It was then that the services started to be seen as an economic good, that is as foundation for industrialization and economic growth.

Autonomous agency

In the early sixties, a new economic development cycle was growing, following the CEPAL (Executive Commission for Latin America) model of import substitution process promoted in many Latin America countries. The new model demanded an urgent provision of public services in urban areas as infrastructure was necessary to start up the plan. The acceleration of the urbanization process which occurred in that period reflected the new capital accumulation pattern where a massive investment in urban infrastructure was part of the strategy for encouraging productive activities.

The urban water supply and sanitation sector had not only to provide basic conditions of living for the labour force but also allow the expansion of many other industries related to the service, particularly construction.

Using the CEPAL model, international organizations like The World Bank and The Inter-American Bank gave loans as part of an agenda where the social problems were addressed and prioritized (Cardoso, 1982). The terms of the Inter-American Bank loan, meant a change in the national approach to urban Watsan. The bank demanded dynamic organizations, independent from political interference and pressures to ensure that the money would be used according to the initial stated purpose.

In addition, it was essential for the Bank to achieve a return on the investment. Tariffs were required to cover investment costs as well as operation and maintenance. The institutional framework set up as a result of those conditions comprised organisations officially attached to the regional government but quite autonomous.

In Salvador, Brazil, the regional government stated that the water supply service would be managed as an industry, as concerned about revenues and profit as any other industry (Magalhaes, 1961). The physical network was mainly directed to the areas with greatest potential for revenue generation. The financing model played a key role in the mode of management. This official proposal, presenting the service as an economic good, represented a significant change from a political leadership who usually promised services to the poor knowing that they could not deliver.

Centralized control

From the 1970s, to the 1990s, the pattern of urban Watsan was established by the military central government as a consequence of the centralized managerial model practised during that time for public policies.

This model was market based (as opposed to welfare oriented) as described by Melo (1989), and Coing et al (1989). Coing explains how it aimed to have a self-financing and profitable Watsan sector. The policy was based on regional state companies using a tariff system based on financial transfers from high-income to low-income householders. The industrial management of the sector remained an important idea.

Discussion

So far the alternative institutional approaches have been seen to be the consequence of external influences or the consequence of failure in the sector. Nevertheless, it is important to analyse the constant dilemma faced by the authorities in low-income countries, where the companies face an endless deficit spiral.

Coing et al (1988) say that the management mode is a result of change in social relations amongst the users as well as in the use of water. In the case studied in this paper, and in many other Latin American countries, the dilemma faced by the public authorities is to balance the contradictions between the different demands from high-income and low-income householders in the same geographical area, a problem that has been constant throughout the 20th century.

Despite the constant official line that the changes would provide efficacy and sustainability, political considerations have never allowed organizations to charge the real cost of water to the customers. Even in the two last changes in the 1960s and 1970s, dictated by important external actors (Inter American Bank and the Central Government), the authorities did not succeed in charging the real costs. The vast majority of public water companies in Brazil and more than half in Latin America are run with large deficits.

The political aspects of setting a tariff according to the real cost is immense in an environment where the poor do not have access to many of their basic needs, and there are large inequalities in society. The price of water is part of the welfare role played by the public authorities. At the same time they play the game set by the technocrats, opting for global technical solutions but not taking into consideration the subsequent feasibility for customers who are unable to pay properly for the service. 'Poor households cannot afford the design standard of industrial countries, but such standards are not necessary on health grounds' (World Bank, 1993:46).

In practice, there is a lack of political will among the authorities, making it difficult to charge the population. It is then said that there is no willingness to pay which encourages the utilization of a large number of subsidies that, in fact, will only provide cheap services, sometimes at a high standard, for the rich minority of the population.

Jacobi (1989) studied the difficulties faced by low-income householders in poor areas of Sao Paulo, Brazil in obtaining water connections. Carteado (1993) quantified the number of complaints by customers lacking water in high-income and low-income areas of Salvador, Brazil during the period 1976-1986 which demonstrated discrimination in water delivery.

The changing policy role played by the public authorities causes the management of the sector to progress in a pendular fashion. Sometimes, the leading conception is that the provision of services is more closely related to economic growth; at other times, it is believed that the service is a right for every citizen, as water is basic for survival. However, even when one of the views is leading, the other still has a role.

This is the dilemma faced by the authorities. When the efficacy thesis for economic development predominates, the institutional approach is defined by self-sustainability and an entrepreneurial managerial. It is said that the service ought to be run as any other industry. Measures should be taken in order to make the activity attractive to the private sector, the market demand is prioritized, the tariffs tend to be set taking into consideration not only the operational and maintenance costs, but also the cost of investment. The project engineering during those periods is tempted to emulate industrialized country solutions, prioritizing the water production and transmission, rather than the distribution network.

When the opposite view is predominant, in which the provision of water is seen as a basic right for every citizen not related to his/her income, government actions lead to extending the physical network towards the peripheral areas. The technological approach then, where possible, produces local and cheaper solutions. There is also a closeness between the activity and public health activities. Within this perspective, the tariff must be related to the presumed 'ability to pay' of the customer, not directly to the cost or quality of service.

INSTITUTIONAL MODELS FOR URBAN WATSAN	PROVIDER		PURCHASER
	Wholesale (Bulk provision)	Retail (Distribution)	

INSTITUTION	*Characteristics*	Wholesale	Retail	Purchaser
GOVERNMENT				
Government/State Ministry	Centralised government control	1970-	1990	Through
Municipal Department *Regie Directe*	Centralised municipal control	1925-	1960	subsidies
Decentralised Government Department *Regie Directe*	Encouraging autonomy at district or municipal level	1960-	1970	
Semi-autonomous Utility Department *Regie Autonome*	Managerial autonomy	1970-	1990	
Autonomous Utility Board *Regie Personnalisee*	Managerial, financial and legal autonomy			
Regulator	Ensures adherence to agreed performance standards including tariffs			
PRIVATE ENTERPRISE				
Services Contractor	Fees per task - conventional contractors	1925-	1990	
Informal Sector or 'Mini' Contractors	Fees per task - small scale enterprises			
Management Contractor *Gerance*	Fees proportional to physical output of fixed assets	1905-	1925	
Shared Profit *Regie Interesse*	Fees proportional to output plus bonus or shared profits			
Leasing Contract *Affermage*	Agreed amount of tariff to lessee of fixed assets			
Concession Contract (BOO/BOT/BOOT) *Concession*	Agreed amount of tariff to concessionaire who has constructed fixed assets	1853-	1905	
Private Equity Owner *Divestiture*	Profit from operations to shareholders who own fixed assets	1995?	1995?	
COMMUNITY				
Community Group	Informal group			
Cooperative	Formal group		1995?	
Non Governmental Organization (NGO)	Voluntary agency providing services			
CONSUMER				
Household	Owner or tenant			1995?
Institution	Government			1995?
Commercial/ Industrial	Private enterprise			1995?

Figure 1. Institutional models for urban Watsan

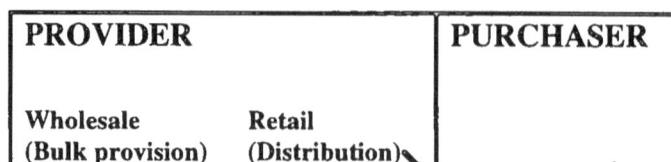

But even then Coing (1992) explains that the possibility of exploiting Watsan services did not work properly because countries did not have the capacity to provide a consumption market that allowed the largest part of the population to become part of the market. Thus, a large part of the population has no access to the service, making it difficult for the country to reinvest within the Watsan sector. In Salvador, even after a large investment in a sewerage system during the 1970s only 18 per cent of the population had proper sanitation services in 1990 (Carteado, 1993).

The future

Whilst the pendulum of institutional change keeps swinging it remains a fact that the poorest do not receive from the public service the water they need for basic needs health. Similarly industry often does not receive the water it needs for economic growth. The poorest end up purchasing water from vendors at many times the cost per cubic metre that they would pay through the official system (in some cases over one hundred times as much) and industry ends up paying more by contructing its own facilities to ensure a reliable supply, but thereby forfeiting the desired economies of scale. It can be argued that the main beneficiaries of the present system are the high and middle-income domestic consumers with household connections and household storage tanks.

Alternative means of managing Watsan services are now being considered again with the emphasis on private sector involvement. An example of the 'wheel coming full circle' again for Bahia? The main options appear to be either a leasing contract or complete divestiture through sale of the equity. Adapting Coyaud (1988), the table lists the options now seen as available for Watsan provision. The table additionally makes the distinction between the wholesale provider and the retail provider. Quite different skills are required for bulk abstraction, treatment and transmission of water (wholesaling) as opposed to distribution, tariff collection and consumer service (retailing). UNICEF (1992) have been involved in an example of a retailer approach in Tegucigalpa, Honduras where 45 000 people in 25 low-income neighbourhoods have received water at less cost than they used to pay to vendors through sale of water from the public network to community tanks.

In considering all the options and choices available to ensure a viable institutional framework for Watsan serv-ices for the whole range of consumers to achieve public health and economic growth it is now necessary to accept that there will be a network of 'actors', involved in both providing and purchasing.

No decisions have yet been taken as to which is the most suitable combination of 'actors' for Bahia. The conclusion is that although the institutional framework plays a key role in the style of managing, the most important idea that will allow countries to overcome the constant deficit crisis in urban Watsan in Latin America will be the political willingness to make the sector feasible not only in speeches but more importantly in practice.

References

Cardoso, F H., (1982). 'Las Politicas Sociales en la decada de 80. Nuevas Opciones?', Simposio Internacional sobre politicas de Desarrollo, Santiago del Chile.

Carteado, Maria de Fatima Dimas. (1993). Servicos de Agua e Esgoto em Salvador: Trajetoria Institucional. Salvador. Brazil. EAUFBa. Unpublished Msc. dissertation.

Coing, Henri. (1992). Les Services Urbains Revisitees. Ecole Nationale des Ponts et Chaussees. Noisy le Grand, France.

Coing, Henri; Lara, Philippe; Montano, Iraida. (1989). Privatisation et Regulation des Services Urbains. Une etude comparative. Ecole Nationale des Ponts et Chaussees. Noisy le Grand. France.

Coyaud D P, (1988), 'Private and public alternatives for providing water supply and sewerage services.' World Bank Infrastructure and Urban Development Report INV 31, Washington.

Jacobi, Pedro, (1989), Movimentos Socias e Politicas Publicas, Sao Paulo, Ed Cortez.

Magalhaes, Juracy. (1961). Mensagem apresentada a Assembleia Legislativa em 7 de Abril de 1961. Salvador, Brazil. Imprensa Oficial.

Melo, Marcus Andre B. de. (1989). O padrao brasileiro de Intervencao Publica no Saneamento Basico. Revista da Administracao Publica, v.23. Rio de Janeiro, Brazil, p.84-102.

UNICEF, (1992), Waterfront, Issue 1, February 1992, New York.

World Bank. World Bank Report 1993. (1993). Oxford University Press, New York.

Application of key sector principles: issues and realities

Rekha Dayal and Peter Lochery, India

GLOBAL CONSENSUS AT the end of the Water and Sanitation Decade on key principles to guide the sector, has resulted in several initiatives to design and implement projects using participatory approaches and a combination of institutional options. In the context of the sociocultural, political and sector policy environment in South Asia, the following three principles pose a particular challenge: (i) demand orientation, using client centred approaches, (ii) treating water as an economic good with inbuilt systems for cost recovery and (iii) management at the lowest appropriate level with implications for strengthening local bodies, NGOs and community-based organization (CBOs).

Based on the emerging experiences from the UNDP-World Bank Regional Water and Sanitation Group for South Asia (RWSG-SA) work program, which supports preparation and implementation of rural and urban water supply and sanitation projects, the authors argue for: (i) use of an adaptive design with mechanisms for frequent internal reviews and corrective actions incorporated in the managerial processes for preparation, implementation and monitoring and evaluation; with the objective of improving project effectiveness by ensuring that accountability and ownership rests with participating organizations and communities; (ii) learning oriented methodologies for implementation, including careful documentation, dissemination and use of lessons learned; and (iii) using intermediation as a strategy for addressing institutional issues and concerns.

Financial resources available from donors and government for water and sanitation projects have always fallen far short of the need. Studies focusing on participation along with emerging experiences from selected projects in the region provide evidence that users are unlikely to finance schemes unless planned facilities meet their needs at a price they can afford. This is more likely to take place in projects in which users participate in decision-making from the very beginning. Financing mechanisms need to be considered in terms of their accessibility to the poor and their cost effectiveness, for recovery of capital costs as well as for provision of recurrent costs. Intermediation by informal institutions in countries like Bangladesh, India and Pakistan may play a role in helping the poor gain access to credit or other opportunities to mobilize the means to pay for the services they want. The key questions being examined are: what is the range of informal institutions and processes involved, and how do they influence the financial stability of projects? What financial level of support is needed for operation and mainte-

nance, and how is this arranged in community based projects which have demonstrated long-term viability.

While client participation is key to designing projects which identify and meet users' self-perceived needs, findings ways to include users in the design and implementation of water supply and sanitation schemes is not easy for formal institutions. Institutions in the informal sector, such as NGOs and CBOs, have been found to provide essential intermediation services by facilitating the participation of clients, especially the poor and particularly poor women, in the design and implementation of projects. Once a project has been constructed, however, many NGOs move on to other communities. The flow of benefits from a project tends to dwindle if operation and maintenance are not managed in a sustainable way. This is found to hinge on the development of institutions which continue to provide intermediation services, or on the development and evolution of self-regulating systems. These are difficult to identify initially and are more likely to surface through an iterative process. Consequently project designs which are programmatic in nature and make provision for the initial project design to be reviewed and amended through internal mechanisms, offer much greater potential for identifying and implementing successful intermediation processes as well as helping to define the factors governing intermediation.

Modest experiences from the region show that where project designs are in the form of blueprints, there is limited potential for addressing these questions through learning from and feeding back experience. Monitoring and evaluation is often perfunctory due to lack of incentive, and changes in implementation, operation and maintenance procedures are hindered by excessive bureaucracy. Initial steps in developing these internal mechanisms are being taken by RWSG-SA through innovative wss projects in Bangladesh, Sri Lanka and Nepal as well as more traditional projects elsewhere in the region. Other regional groups of the UNDP-World Bank Water and Sanitation Program are also doing similar work. The Program is in the process of synthesizing global experiences.

The process of learning through an ongoing project has been found to require flexibility in project design and also needs to be adequately structured and resourced. Project staff are often under considerable pressure to provide short-term quantitative results and need a framework to assist them in making objective procedural changes. The frameworks developed to date have consisted of series of hypotheses, which the 'structured' learning is designed to validate. Where initial hypotheses are found to be flawed, appropriate changes in implementation, operation and maintenance can be made.

In addition to using conventional forms of dissemination of materials, the Program promotes inter-regional exchange of experiences through workshops. One such forthcoming event is a regional workshop for preparing and implementing large scale rural water supply and sanitation projects in South and East Asia, for project managers of World Bank-financed projects. The workshop will provide a forum to share experiences, problems and solutions, to establish a support network for information and resource sharing, and to encourage the adoption of a systematic learning and documentation approach in ongoing projects.

Rekha Dayal is Regional Program Advisor and Peter Lochery is Regional Manager for South Asia of the UNDP - World Bank Water and Sanitation Program.

Water and economic development

Dr. V.J. Emmanuel, Sri Lanka

THE TOTALITY OF human endeavour is directed to securing the highest possible level of welfare and comfort for man. Every advancement of science and technology has this for its ultimate aim. The process of applying technical knowledge to the art of living is generally described by the term 'development'.

This presentation has for its objective the re-orientation of the approach of the water engineer to the subject of water engineering as an applied science, and the introduction of the concept of his participation in total economic development. The inter-relationship of the disciplines of economics and economic development, philosophy, psychology, physiology, sociology and the behavioural sciences become self-apparent.

Man, like all other living organisms, is intimately linked with and influenced by the environment which surrounds him, and his life and functions are controlled by the suitability or otherwise of the environment.

The environment

The concept of preventive medicine has changed considerably in recent times, and is now being directed more and more towards the study of the relationship between man and his environment. The elimination of diseases caused by an unfavourable environment can only be secured by an adjustment between man and the environment. This adjustment can be achieved in two ways; either by man altering his way of living to suit the environment, or by the modification of the environment to suit man's way of life. The final answer will, no doubt, lie in a combination of these two approaches.

The conditions under which diseases of an unfavourable environment cannot survive can be brought about only by co-ordinated progress in the economic, cultural and social spheres, leading to a better pattern of living. Since this process is of an evolutionary nature, leading to a better pattern of living, it must be watched and guided if the final result is to an integrated whole.

Physical environment and in particular, 'sanitary' environment, has for a long time been recognized as having a profound influence on health. However, it is only comparatively recently that the importance of the social environment in regard to health has received recognition.

Water and health

Dr. Candau, one-time director general of the World Health Organization, in his address to the International Conference on Water for Peace, held in the U.S.A. in 1967, stated:

'Man can live without clothes, without shelter, and for some time, without food; without water, however, he soon perishes. It is not surprising, therefore, that for the early periods of man's history, his dwelling places have been closely associated with lakes, rivers, springs and wells. Civilization and cultures were nurtured in the valleys of the great rivers - the Nile, the Indus, the Ganges and the Yangtse'.

Dr. William H. Stewart, Surgeon General, USPHS, addressing the same conference said: 'No resource is more basic to health progress than water — water that is safe to drink and cook with, water that is easily accessible in sufficient quantities. It is true that man cannot live without water; but it is also true that in much of the world, the water upon which he is dependant is itself a threat to life. In the developing nations, insanitary water is still a carrier of diseases like cholera and typhoid fever and other parasitic infections that drain energy and hope from millions of people. Unclean water is the source of diarrhoeal diseases that claim the lives of over five million infants a year, before their first birthday'.

The physiology and psychology of entrepreneurship

Several social scientists and economists have expressed the view that people in developing countries, particularly the Asians, have low productivity because they are 'naturally' lazy and undisciplined, and possibly have a lower IQ than their Western counterparts. This view was extended to mean that most of the developing countries could not be expected to reach the level of development or intellectual attainment of Western countries.

This canard has already been disproved by the performance of people from developing countries in the Western environment, as also by the intellectual, academic, professional and industrial attainments of a section of the people in these backward countries.

There is no gainsaying the fact, however, that the majority of the population in the developing countries have very low productivity and lack the initiative, enthusiasm and drive to improve themselves. Their philosophy seems to be: 'Sufficient unto the day is the evil thereof'. The reasons for the apparent sluggish economy lies elsewhere.

Most of the developing countries are still 85 per cent rural, with an agriculture-based economy. The rural population do not own land, and are either tenant farmers or seek employment in agriculture. These people have a low basic health level, and are virtually at 'survival' level. The majority of the rural population suffers from water- and filth-borne diseases, many of them from multiple infections. The continued low level of basic health in these areas, attributable to diseases transmitted through contaminated water or soil, is perpetuated by the lack of safe and adequate water and basic sanitation.

Water-borne diseases and soil transmitted diseases (particularly the helminths Ascaris and Ankylostoma), sap the energy and zest for life, and generate a distaste for any avoidable exertion. Then drain all ambition, drive, initiative and the desire to strive for self-improvement.

A high level of energy in the individual is a prerequisite for the ambition and drive that leads to self-improvement. The urge for self-improvement is the seed that germinates and grows into the spreading tree of diverse achievement. Taken collectively, it leads to economic development. The rate of development in a rural community which is marginally above the 'survival' level, is inescapably linked to the basic health level of the community.

Human welfare has been divided into five levels of concern. These are:

- Survival
- Protection from disease
- Performance
- Enjoyment
- Creativity

The sluggish economic improvement that is evident in a rural community with a low basic health level, can reach the 'take-off' point in self-motivated economic development, only when the stage of 'protection from disease' has been passed.

Water, while supporting life itself, is perhaps the greatest destroyer of life. Water provided for domestic use, should be viewed not from the hydraulic aspects alone, but from the public health aspect as well. Economic development can only ride on the back of a sound basic health level of the community, and the basic health level is tied up with the water the community drinks and washes with. Domestic-use water, it must be remembered, has to be both safe and adequate to support a high level of personal hygiene.

If domestic-use water is to be provided to the rural population which is scattered in small groups, and if it is the intention of the local authority that the system must be operated and maintained by the beneficiaries, it must necessarily be through a system which is relatively inexpensive to construct, and simple to operate and maintain. The logic for 'AFFORDABILITY' is self-evident.

Man — the primary resource

Man is the primary resource in economic development. The mental attitude of improvement, motivation, ambition, drive and perseverance are essential to economic development.

Several economists have described economics and economic activity as consisting of the constant struggle on the part of man to subdue nature to satisfy his needs. This theory defines two categories of resources — the human agent, with all his powers of brain and brawn, emotions and skills; and the external physical world which he tames to his own purpose. The interaction between man and his physical environment is the area of economic activity.

If man has a fundamental role in economic development, and freedom from diseases of the environment is essential to his having the mental attitude or motivation for development, then without doubt, the water engineer has a key role in producing the climate for economic growth.

Economic development and health

Water engineers are well aware of the strong influence safe water has on human health. WHO has defined health as a 'state of complete physical, mental and social well-being, and not merely the absence of disease or infirmity'. For practical purposes health can be appraised by 'the extent to which the human body is capable of fulfilling its physical and mental functions, producing effective work and of enjoying life in given conditions — genetic and environmental'.

Ben Abbot made a clear analysis of the influence of man's health on his mental and physical outlook. 'Poor health', he stated, 'causes adverse conditions in varying degrees, such as the following:

- the mind is not clear and alert;
- he cannot think logically. The thinking process is slow. Reasoning power and judgement are affected;
- the mind is more apt to give in to pessimistic, discouraging moods, with greater emphasis on CAN'T than on CAN
- it is more difficult to concentrate, make decisions and persevere in undertakings to successful conclusions;
- initiative is crippled. The habit of taking action is undermined;
- production and efficiency are materially reduced;
- resistance to fears and bad habits is lowered;
- emotions are harder to control and one is more apt to be timid and diffident;
- time is seldom used to best advantage;
- interest and enthusiasm are lessened;
- it becomes much harder to overcome difficulties and start action after mistakes;
- nervous disorders and illnesses are more prevalent;
- self-respect is diminished and it is harder to work with and influence others".

Time and again, member countries of WHO have clearly cited unsafe water and water shortage for the continuing low social and economic development in their communities. In 1959 the World Health Assembly, 'recognizing that safe and adequate measures for the protection and improvement of health are indispensable for economic and social development', launched a global water supply programme.

Engineering technology and economics

Economics is part and parcel of the science of engineering. Broadly speaking, engineering is a science which can produce a given result, at highest efficiency, at the lowest cost. All water engineers should therefore be concerned with the principles of economics, and society places on them the obligation to conceive and construct schemes which are economical and which will produce the highest benefits in relation to money spent. Each individual scheme should be subjected to the tests commonly applied in economics, so that the best alternative is selected to serve the purpose in mind. In such an evaluation, the sophistication of the scheme should be tailored to the social and economic status of the community to be served; it should be affordable and sustainable by the community. It has to be borne in mind that to serve the health objectives, a high level of sophistication in project concept in not necessarily called for. This does not mean that a simple project does not call for sophisticated technology. A good case in point is the technology required in location and drilling deep wells in fissured hard rock areas. Affordability dictates that deep wells, furnished with hand pumps, will be the most appropriate solution to providing safe water in a given community. These rural people had traditionally to walk several kilometres to polluted ponds, to fetch their minimum domestic water requirements. Drilled wells drawing water from deep fissures in the rock providing safe and plentiful water at their doorstep, were a sustainable and adequate solution. Haphazard location of deep, drilled wells in hard rock areas, with a view to intersecting the water-bearing fissures at a depth adequate to provide a perennial supply, could turn out to be a grossly prohibitive venture, thus negating the purpose of affordability. The technology for locating water in the fissured rock, at a depth which would provide perennial water; pin-pointing the exact intersection at ground level, and drilling to depths of 30 to 70 metres, involved Resistivity and Seismic Refraction technology, with computer linked data interpretation, plus skill in the operation and maintenance of DTH-hammer rigs. The end result was a simple drilled well and handpump, which was well within the affordability and sustainability of the rural population.

Engineers are prone to be grandiose in their conceptualization, and it is in the nature of their training and skill to prefer sophisticated and elaborate solutions, even where the community finances cannot support them. This tendency is evident in the field of water supply, because water engineers propose elaborate and expensive schemes to serve simple needs, at once putting such a programme outside the financial resources of the community and the country.

The pragmatic approach

In developing countries which are largely agricultural and 85 per cent rural, and where the ability of the community to pay for services is low, there is a strong need for low-cost water. The water engineer in a developing country is therefore faced with the challenge of devising simple, inexpensive methods of providing rural communities with their basic safe water requirements. This will require that engineers come to grips with reality, 'because a modest programme that can be executed is to be greatly preferred to an elaborate one that never gets off the paper'. Water engineers must resist the temptation to associate in their minds, the science of water engineering only with developed urban communities and sophisticated schemes; for in countries where 85 per cent of the population is rural, this will in itself constitute an unrealistic attitude.

Social and economic development

Spangler, in his comments on John Mills' exposé on Economic Development, states:

'... economic progress depended, as did the augmentation of human welfare, upon two sorts of improvement —the extension of man's knowledge of the laws of nature and his capacity to remove barriers imposed by an unbeneficient nature; and upon the removal of barriers imposed by men on themselves (in the form of beliefs, customs, opinions and habits of thought), together with the sustenance of forces that made man strive to improve and elevate human nature and life'. The study between cultural patterns and physical conditions is of the greatest importance for 'an understanding of human society, but it cannot be undertaken in terms of simple geographical controls alleged to be identifiable on sight'.

The magnitude and urgency of the problem

The World Health Assembly, by a resolution in 1980, set in motion a programme with the objective of providing safe water for everyone by the year 1990. While some momentum was created, achievement fell far short of the target. The 'Water Decade' was extended to the year 2000, and time is rapidly running out. Some studies reveal that the annual population coverage falls short, even of the annual population increase in those countries. This also takes into account systems which fall into disuse due to lack of maintenance or sustainability. Granted that the human attitude for self-improvement can be generated only when a basic health level is achieved, water engineers may have to review their approach to providing 'safe water' as a step-wise process, achieving the health objectives at an affordable cost in the first instance, and upgrading the system as the affordability and expectations of the community rises. The challenge to water engineers lies in using sophisticated knowledge and technology to find simple solutions to safe and adequate water supplies, within the affordability of the community being served.

References

Candau, M.G., *Water for Living*, International Conference on Water for Peace, 1967.

Stewart, W.H., *Education and Welfare*, International Conference on Water for Peace, 1967.

Abbott, B., *The Spur to Progress*, Mind and Work Series 9.

Spangler, J.J., *John Stewart on Economic Development*, Free Press, New York.

World Bank (IBRD), *Water and Economic Development*, International Conference on Water for Peace, 1967.

Enke, S., *Economics for Development*, Prentice-Hall Inc., 1967.

Institutional constraints in the Maharashtra water sector

Mark Harvey, ODA (UK), Bombay, India

THE GOVERNMENT OF MAHARASHTRA (GoM) state in west India is currently implementing a rural water supply and sanitation project supported by the Overseas Development Administration (ODA) of the UK.

Institutional difficulties have emerged at various times during the life of this project. Some of these difficulties are apparent from asking straightforward questions such as 'who is responsible for Operation and Maintenance (O&M)?'. Other difficulties are hidden more deeply beneath apparently solvable technical problems.

These institutional complexities are illustrated in this paper by looking at source development problems that have occurred on two regional schemes. Two different technical problems, in engineering terms, are analysed to identify institutional or sectoral problems and solutions.

Maharashtra state

Maharashtra state is the most developed of Indian states in terms of mobilization of surface water resources through large dams. However, this utilization of available water has a significant impact on users of water other than those of irrigated agriculture, who account for over 95 per cent of the usage of surface water storages. This impact is largely felt by users of 'small' quantities of water whose access to the substantial resources, that are locked up by large dams, is limited.

The water sector in Maharashtra is institutionally complex. There is no over-arching state-level water resources agency, although the Irrigation Department consider themselves as such as far as surface water resources are concerned. Neither is there a state water policy, although by default national water policy is followed.

The Maharashtra project

The Maharashtra Rural Water Supply and Sanitation Project (MRWSSP) supported by ODA covers four regional (ie multi-village) piped water schemes in three districts.

The project's lead agency is the Rural Development Department (RDD) and the project aims to develop an integrated approach to the provision of drinking water systems to 200 villages. This approach is to be achieved by linking water supply engineering, managed by the Maharashtra Water Supply and Sewerage Board (MWSSB); with health education (HE) activities, managed by the Public Health Department (PHD); and community participation (CP) initiatives managed through the district councils (Zilla Parishads) (ZPs). The ZPs are

administratively under the RDD. They are supported in their CP activities by community development consultants from the Women's Studies Unit of the Tata Institute of Social Sciences (WSU-TISS).

Without dwelling on the institutional difficulties of integrating the objectives and activities of the five agencies mentioned above the project can be best described as being born of an older generation. It is an engineering project that has had HE and CP components added to it.

The water supply engineering component consists of four large schemes of 80, 51, 22 and 56 village groups. Two of the schemes have existing reservoirs as their source, the other two are based on run-of-river sources. All schemes lift water to treatment plants and distribute it by gravity mains and occasional booster pumping to elevated service reservoirs and internal distribution systems in each village.

Currently the 51 and 22 village schemes are suspended due to water availability problems at their run-of-river source. Last year the 56 village scheme reservoir intake construction was delayed for the third year running due to flooding. It is these technical problems at the sources that are analysed below to demonstrate the institutional constraints to effective water resource management.

Source problem 1 — The 51 village scheme

The Tapi River was a perennial river some years ago. However it now has insufficient flow during the late dry season to sustain the source for this scheme. This problem became apparent to MWSSB during 1991, although construction work had started on some elements of the scheme prior to 1990. The Tapi is a notified river and so comes under the control of the state Irrigation Department (ID). MWSSB therefore applied to the Irrigation Department for permission to lift 3.6 Mld for this scheme. In response the ID advised that this was acceptable but that four months storage had to be provided to cater for the period February-May when river flows could not be 'guaranteed'. The period of four months was defined by the ID. Work done elsewhere (Vincent, 1993) suggests that things are not as simple as this, see Table 1. Not only does the water have to be available (a minimum of 3.6 Mld or 42 l/s) but it has to be capable of being drawn. Shallow flows or those which meander from the intake are of no help even if sufficient in quantity.

From January 1992 to the present time MWSSB have been attempting to find a viable solution to this problem. They have had assistance from ODA, the state Groundwater Survey and Development Agency (GSDA), the ID,

Table 1: Periods of low flow at Sukwad gauging station, Tapi River

Year	< 10 m3/s	< 2 m3/s	< 1 m3/s	< 0.5 m3/s
1985/86	15 Nov – 16 Jun	12 Apr – 16 Jun	1 Jun – 16 Jun	1 Jun – 16 Jun
1986/87	1 Oct – 17 Jun 11 Jul – 16 Jul	9 Feb – 17 Jun	15 Apr – 31 May	28 Apr – 31 May
1987/88	15 Sep – 13 Jun	17 Jan – 10 Apr 1 Jun – 13 Jun	1 Feb – 16 Mar 4 May – 15 May 1 Jun – 13 Jun	1 Jun – 13 Jun
1988/89	1 Dec – 16 Mar 1 Apr – 31 May 1 Jun – 18 Jun	1 Mar – 23 Mar 15 Apr – 6 May 30 May – 31 May	27 Apr – 6 May	
1989/90	28 Feb – 1 Apr 13 Apr – 17 May 1 Jun – 12 Jun	6 May – 14 May 1 Jun – 12 Jun	6 May – 14 May	6 May – 14 May

consultants (both local and from UK) and local communities (by providing local knowledge).

It is now emerging that the most likely solution to MWSSB's problem is one identified by local communities. It took some time for MWSSB and others to be alert to this local knowledge.

The problem can be defined as one of hydrological information and water resource management. However other factors have had a bearing and these are discussed below. It should not be forgotten that different stakeholders would define the problem in different ways. There are institutional limits to the ways in which a problem will be defined. It follows therefore that there can be institutional solutions to such problems.

Source problem 2 — the 56 village scheme

The existing Girna irrigation reservoir is the source for this scheme. There is no problem of water availability here, the problem was one of 'too much' water which caused construction delays at the intake.

Construction of the headworks (intake) started in 1990. During the monsoons of 1990, 1991 and 1992 the construction site was inundated by rising water levels in the reservoir. In fact, in 1991 and 1992 the water levels did not fall sufficiently low, for a long enough period of time in the later months of the dry season (January-May), to allow any substantial construction work to be undertaken.

Figure 1 shows a plot of reservoir level against time over the past few years since the Girna Dam was impounded. Key construction levels are shown. The plot shows that a low-level cofferdam will only allow short periods for construction. The risk taken by the contractor was to build such a cofferdam to a level high enough to allow the works to be constructed but low enough to keep costs to a minimum. The gamble failed in 1991 and 1992.

The risk taken by the contractor was understandable when the contractor is only likely to win the work by tendering the lowest cost. Since this time MWSSB have considered developing procedures for pre-qualification of contractors where works are particularly difficult.

Eventually advice was sought by MWSSB from the ID regarding the height and safety aspects of the cofferdam. In response the ID recommended construction of a cofferdam to the same level as the crest of the Girna Dam itself, with substantial zoning and clay core material in its structure, enough to make it worthy of a permanent earth embankment. Such a proposal was the lowest risk option, very cautionary, but prohibitively costly as far as the contractor was concerned.

In 1993 the contractor gambled again, and won. A new, 'medium' height, cofferdam was built, MWSSB revised some construction details and good progress on site enabled permanent works to reach a level above any possible water levels during that year's wet season.

Unlike the 51 village scheme the problem here has been solved. It was essentially a site investigation and construction management problem. But, as stated above different stakeholders would define the problem in different ways. The contractor might define it as a problem of natural causes that was outside his capacity to solve; MWSSB might define it as a problem of contractors' tendering and appointment procedures (lowest cost = least experienced ?); the donor might define it as a problem of poor management or poor design.

Problems encountered — engineering or institutional?

The two problems described above are different — one was a problem of insufficient water, the other was one of abundant water. The two problems can also be considered as similar, both having their roots in inadequate source assessment in its broadest sense; that is, source assessment for security, quality, quantity, time, place, construction, operation, legality, etc.

If the problems are considered as being different technically, in engineering terms, but at another level having the same root cause, then this proposition can be analysed further to identify institutional or sectoral problems and solutions.

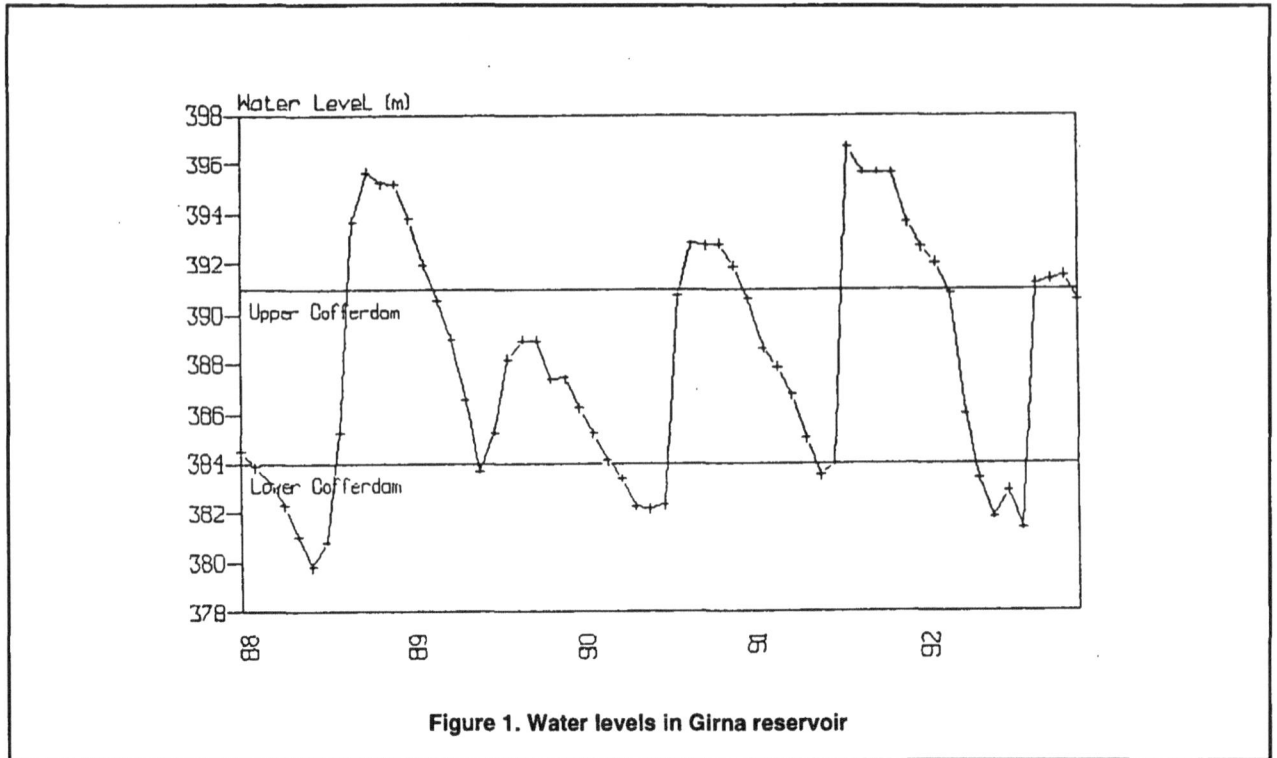

Figure 1. Water levels in Girna reservoir

Institutional analysis

The MWSSB was created as a semi-autonomous board in 1979. It was created out of the GoM Public Health Engineering Department (PHED) which came under the authority of the Urban Development Department (UDD) at that time.

The MWSSB is staffed professionally by public health engineers with their main activity being domestic water supply. The main scope for gaining different experiences is to move between urban and rural work. Occasionally senior staff have been deputed to the Maharashtra Pollution Control Board (MPCB).

In a similar way engineers of the Irrigation Department do not move outside major, medium and minor projects work. Different roles between those involved with resource mobilization and command area development do exist.

On the emergence of the problems described above, good communication and co-operation was required between these two agencies who have different views of the water resource, different needs and different available skills and experience.

The ID are primarily concerned with storing as much water as possible by the end of the monsoon period and releasing it in two main crop growing periods. Their water planning is in terms of large volumes of water (millions of cubic metres) in blocks of releases at intermittent times.

On the 51 village scheme problem of water availability, the need for a continuous flow at the source of about 1 cumec is very small compared with agricultural demands.

The matter for the ID is perhaps one of routine administration of a small abstraction request whereas for the MWSSB it is the critical need. It is therefore not surprising that the ID's proposition that MWSSB provide four months storage was superficially dealt with by a statement on the abstraction permit without the full consequences being considered.

On the 56 village scheme problem of inundation of works, the need for good analysis and understanding of reservoir levels during the drawing down period was essential. This is essentially the opposite of the ID's main concern which is about how full the reservoir is and how much water is available for use in the next growing season.

The ID dam projects are currently moving from planning dam projects based on 75 per cent reliability to 50 per cent reliability as there are fewer good sites available. This has very significant consequences for domestic water supplies which need 95 per cent reliability. There is a conflict of interests between the ID, who 'control' large surface water sources, and domestic water suppliers who need relatively small but regular flows.

In both cases described above the divergent needs of MWSSB and the ID have been simplified but they are very real. They appear again in the ID's response to the design of a temporary cofferdam; while a cautionary solution was provided it had limited practical application in the circumstances.

A similar divergence of data needs has recently emerged at the 80 village scheme source in the Hatnur Reservoir on the Tapi River. The Hatnur Dam reservoir levels have only been recorded in many years during the monsoon months, when it is filling. Few levels have been recorded since impoundment for the dry summer months. The security of this source now requires further checking.

Lessons to be learned

The main lesson to be learned from these experiences is that the various agencies involved with water in the state — ID, MWSSB, GSDA, MPCB, ZPs, PHD (water quality) and Maharashtra Industrial Development Corporation (MIDC) — need to increase their awareness of the resource needs, data needs and uses that other agencies have for water.

The solution to increasing this awareness between agencies may lie in training, regulations, information exchange and career development, or combinations of these. It can start on occasions such as the WEDC seminar.

Many people call for a river basin management approach to water resources: the author would support this but appreciates that it may not be a simple solution. Maharashtra does not need another layer of bureaucracy that might centralize data and information. Initiatives need to be taken that promote skill development and integrated information resources at local level, where the needs are.

In the interim MWSSB could consider setting up a 'Source Assessment Cell' along the lines of their recently created Quality Control Cell. Such a specialist cell would be called upon to assess source adequacy and security or at least be responsible for co-ordinating the work of other specialists. It is understood that as a result of the difficulties encountered on the 51 village scheme, in the World Bank-supported project in the state, intakes are being limited to existing surface water bodies; no run-of-river intakes are to be used.

The Irrigation Department should be encouraged to gather and store data in SI units conforming to international norms and the needs of other users. They should also ensure that adequate data sets are maintained that cover periods of the year or regimes of flow that may differ from the ID's main needs but are likely to be of value to other water users.

Further institutional difficulties and institutional development

The above analysis has concentrated on one aspect of co-ordination and co-operation between organizations in the water sector in Maharashtra that could be improved. Others that could be discussed in a similar way include water quality monitoring, operation and maintenance and the overall planning process for rural water supply.

However, the difficulties that have arisen provide a wealth of information and experience that must not be forgotten. Improving the institutional memory is an important responsibility that all of us have (Kletz, 1993). This can be done by spreading the message on occasions such as the WEDC Conference, discussing the issues amongst colleagues, remembering the messages and creating comprehensive data bases of reports of problems in site investigation, design, construction and operation and maintenance.

By improving the institutional memory, the institutional environment can be analysed and understood more readily and in turn this will lead to sound institutional development taking place — an essential ingredient in producing well designed projects and sustainable development.

The costs of not developing the institutions, not making information more freely available and not making institutions and professionals accountable for their performance will be serious for Maharashtra towards the turn of the century.

Institutional constraints must be relaxed if water resource constraints are not to have severe consequences for the people of Maharashtra.

References

[1] Kletz, Trevor; 'Lessons From Disaster — How Organisations Have No Memory and Accidents Recur'; Institution of Chemical Engineers; Rugby; 1993.

[2] Vincent, Linden; 'Water Resources Availability and Water Development Institutions in Maharashtra State'; unpublished project consultant's report; Overseas Development Institute; 1993.

Demand creation and affordable sanitation and water

Derrick Owen Ikin, SKAT, Switzerland

IN THIS PAPER the commercial market approach and a social marketing approach to water and sanitation will be considered, keeping in mind the overall aim of achieving a positive health impact in the communities concerned. The problems related to latrine coverage, behavioural changes linked to use of latrines and the use and availability of safe drinking water will also be briefly examined.

The following experiences are based mainly on rural water and sanitation projects in Bangladesh, Indonesia, Lesotho and Namibia. They include input from the WHO-co-ordinated working groups of the Water Supply and Sanitation Collaborative Council, for applied research (1993) and for hygiene and sanitation promotion (1994).

Health impact?

The overall objective of most water and sanitation projects is to improve the health of the people concerned in a sustainable way. However, the findings of a regional informal consultation for South East Asian countries perhaps reflects a worldwide phenomenon in that 'little or no effect on behaviour or health status has been demonstrated' (New Directions for Hygiene and Sanitation Promotion; WHO, 1993). Usually the underlying assumption was that the availability of safe drinking water would lead to improved health as is indicated by the numerous projects that focused mainly on water system installation. Sanitation, encouraging hygienic behaviour and health education were and still are given a comparatively low priority. Often safe drinking water is available but sanitation coverage is minimal with unaffordable (for the majority) products being promoted at subsidized rates or even given away free of charge. The users are not usually perceived as customers who respond to choices and if convinced, would use their own resources to meet genuine needs.

Latrine coverage and affordable products

It is considered that latrine coverage has to reach about 90 per cent to have any impact on community health. This is a simplification as obviously behavioural changes and safe drinking water are major factors as well. On a practical level this sanitation 'neglect' translates into hardware-oriented projects, promoting unaffordable products or spot coverage as is often seen in some African villages where two to three VIP (ventilated improved pit) latrines per village are installed and the rest of the villagers use the surrounding areas. A Kenyan friend told of her experiences in finding locked VIP latrines and even one that was used as a storeroom! Regarding sanitation options in both Africa and Asia, the real low cost options are

seldom promoted or are considered inferior to the technologically 'superior' versions. Low cost in this sense is what the majority of people want to and can afford. Spot coverage, however, does have an important seeding role, i.e. the promotion of a status product that people would like to acquire. In trying to understand the complex relationship between creating demand and its link to willingness to pay within the context of product range and availability, we will now briefly examine a rural water and sanitation project in Bangladesh.

Rural water and sanitation, Bangladesh

Water

The Government of Bangladesh/UNICEF rural water supply and sanitation programme has been running for more than 18 years and over 100m US$ have been invested mainly in making safe drinking water available. Today the safe drinking water coverage results are very impressive; over 85 million people have access to safe drinking water (donors: DANIDA and Swiss Development Cooperation). Numerous surveys (WHO and UNICEF) have ascertained that over 90 per cent of the installed pumps are still operating and are thus being repaired and maintained by the users. The policy of involving local pump producers and eventually the stopping of government-funded spares and maintenance fortunately resulted in the users taking over this responsibility. When I first worked in Bangladesh in 1983, tube-well water was considered not fit to drink in rural areas and many wells fell into disrepair. This perception has changed and today most people prefer to drink tube-well water. Thus a need for safe drinking water has been internalized resulting in a willingness to take action to ensure the supply. Over the years a demand has been created and this has resulted in villagers taking action to maintain the pumps and in many cases purchasing their own pumps. The basic hand suction pumps are also relatively inexpensive, and both pumps and spares are now available countrywide. **However, despite this considerable success, it appears that no measurable health impact has been achieved.** This in terms of human suffering is tremendous (countless child deaths due to diarrhoeal diseases) and does not fully utilize the 100m US$ investment made so far.

Sanitation

The rural sanitation aspect of the above programme resulted in the setting up of 1 000 village sanitation centres

(VSCs) where subsidized waterseal latrines and pit rings were sold, often linked as a condition to handpump installation. It is estimated that latrine coverage has increased from below 3 per cent to above 26 per cent (including home-made latrines—HML). However, about 80 per cent of the population cannot even afford the subsidized waterseal latrines promoted by this programme and most NGOs. The dilemma that the government and/or the donors cannot afford to supply latrines to everyone (even if this were advisable), has led to a re-thinking of the long-term concepts of the programme as well as of the roles of private sector and government in sanitation. In the book *Promotion of Rural Sanitation in Bangladesh with Private Sector Participation* (Chadha and Strauss, 1991), the role of the private sector, customer preferences and willingness to pay were examined. This encouraged the realization that the present products are unaffordable to the vast majority of rural people and that the government could never afford to 'provide' so many latrines.

In 1992 while investigating the role of the private latrine producers (PPs) with Professor Mustafizur Rahman of the Dhaka University, we interviewed over 30 latrine producers in rural areas (Ikin and Rahman, 1992). These producers concentrated on low cost waterseal slabs and pit rings as did the government VSCs. The producers sold in competition to the subsidized village sanitation centre latrines. Their products were similar although both sides claimed to produce the better quality! The profit made on latrines is marginal and most producers had to sell other products to survive. However, they often provided a range of options to their customers such as pit digging services, superstructure building, transportation and quality to suit the customers. In a few cases they also provided limited credit. From the producers we learnt that some NGOs, most notably the Grameen Bank, had stopped its own latrine-manufacturing programme and assisted its members to buy from local producers at non-subsidized rates. This encouraged producers enormously and also resulted in improved quality. It is considered that the subsidized latrines and the setting up of village sanitation centres has had a seeding role in promoting latrines. In many cases private producers of latrines have sprung up near the government village sanitation centres and this should increase as the government is now closing old VSCs.

Pioneering promotion linked to a range of latrine options

The Government of Bangladesh (Department of Public Health and Engineering) and UNICEF have started a pioneering social mobilization programme based on home-made latrines consisting of a pit, a bamboo or wooden platform and a superstructure built according to what the users could afford. In the late 1980s this latrine promotion programme resulted in the inhabitants of Banaripara (Barisal district) building over 36 000 home-made latrines with coverage reaching 80 per cent. During programme evaluation/review missions I visited this area in 1991 and 1994. Hanging latrines, usually over a water source, had been virtually eliminated and it was observed in 1994 that the latrines were still being used, repaired and maintained when required. UNICEF has subsequently developed booklets and brochures on building sanitary home-made latrines (Refer *Barisal - A Successful Approach for Sanitation*, UNICEF, 1993.)

Major concerns were and still are the sustainability of the results achieved through social mobilization efforts. This dilemma is discussed in 'Social Mobilization and Social Marketing in Developing Communities' (McKee, 1992) and the main concerns in this programme relate to behaviour changes and health impact. To deal with these concerns extensive promotion material has been produced that focuses on the basic hygiene behavioural changes that should ensure health improvement. The safe disposal of faeces is seen as a first step in this direction. In addition, handwashing and the safe disposal of children's faeces are strongly encouraged. A survey of the social mobilization areas is planned for 1994 and this should indicate improvement in the existing programme and if any health impact has been achieved.

It seems that the promotion of home-made latrines by prominent people, schools and leading local people has increased their acceptability. The focus on privacy, and the links to religion and hygiene, are also factors used in the promotion of home-made latrines. The programme has incorporated many experiences into the more recent versions of social mobilization, but ongoing research and monitoring are needed to establish (a) if any long-term health impact has been achieved; and (b) if the behavioural changes are sustainable from generation to generation.

The programme also encourages buying from private producers (PPs) where the users can afford the more expensive latrines and it was observed that in the areas where social mobilization had already taken place, PPs have set up businesses to meet the increased demand. Although it is premature to judge this programme, so far the indications are that a certain degree of success has been achieved. As seen from a marketing point of view, this success can be summarized as follows:

(a) Mobilizing the population to take affordable action is a way of creating a real demand that together with other factors such as peer pressure, results in action i.e. building sanitary latrines.

(b) Demand creation worked because it was linked (through social mobilization) to an affordable range of latrines.

(c) A known technology (home-made latrines) was made acceptable through promotion by status people and organisations.

(d) Commercial latrine producers and latrine pit cleaners report increased business.

Despite the initial success, the risks are still there and are as follows:

○ Home-made latrines have not been tested in the long-term.

• The question of achieving community health impact remains.

○ Establishing sustainable hygiene behavioural changes that will pass on from generation to generation.

○ Neglecting the need to repeat social mobilization to ensure long-term benefits.

At present the Government of Bangladesh and UNICEF are undertaking numerous surveys and activities to establish among other things:

○ The number of latrine producers in Bangladesh.

• The market situation and consumer preferences in areas where village sanitation centres have been established for a number of years.

○ The introduction of sanplats and the investigation of other alternative low cost latrines.

The results will provide input for drawing up concepts and policies for the next programme phase.

Conclusions

Demand creation, linked to a range of affordable products, is part of the key to sustainability and as such should play an increasing role in programme design and national water and sanitation policy. This involves treating users as customers who are seen as a market for water and sanitation products. Successful demand creation is linked to willingness to pay as is shown by the Bangladeshi villagers who buy pump spares, maintain the pumps and also build sanitary home-made latrines using valuable/saleable home-grown materials.

The benefits of working with and actively encouraging commercial producers are shown in the Bangladesh programme. Water pump manufacturers have developed as a result of large purchases by the programme (not without problems). This has resulted in producers not attached to the programme setting up production and the market taking over the distribution and sales of spare parts and pumps.

Obviously a market-only approach is not possible or in many cases not advisable, but an exclusively service-oriented programme is also unlikely to solve long-term problems. Changing the momentum of a water and sanitation programme can be most difficult, particularly when many programme staff and NGOs are fixated on a particular product or technology, be it a waterseal slab or an African VIP latrine, despite problems and viable alternatives being available. This is further complicated by having to change staffing to suit a new programme focus, i.e. promotion, education or moving from a focus on water to a more balanced water and sanitation mix.

Considering the continuing suffering and dying from sanitation-related diseases as well as the relatively large worldwide investment in water systems and to a far lesser extent in sanitation, it is apparent that new approaches need to be urgently tried and applied. Water and sanitation programmes have already started trying alternative approaches and utilizing these experiences will be the challenge for the next decade.

References

Cairncross, S. 1992. 'Sanitation and water supply: Practical lessons from the decade.' Water and Sanitation Discussion Paper Series No. 9. Washington: World Bank.

Chadha, S. & Strauss, M. 1991. 'Promotion of rural sanitation in Bangladesh with private sector participation.' Dhaka: Swiss Development Cooperation.

Hoque, B.A. & Briend, A. 1991. 'A comparision of local handwashing agents in Bangladesh.' *Journal of Tropical Medicine and Hygiene 94*, 61-64.

Ikin, D.O. & Rahman, M. 1992. Report of the concept mission to develop a proposal for the promotion of latrine production by private producers in rural/semi-rural areas of Bangladesh. St. Gallen: SKAT.

Ikin, D.O. & Walther, P. 1992. Appropriate technology and sustainability for low cost sanitation in the UNICEF/DPHE Rural Water Supply and Sanitation Programme (RWSSP). St. Gallen: SKAT.

Kotler, P. 1988. *Marketing management - Analysis, planning, implementation, and control.* Englewood Cliffs: Prentice-Hall.

McKee, N. 1992. *Social mobilization and social marketing in developing communities - Lessons for communicators.* Penang: Southbound.

Minnatullah, K.M. et al. 1991. 'Community management of rural water supply and sanitation services - A Bangladesh case study.' Dhaka: UNDP-World Bank Regional Water and Sanitation Group South Asia.

OECD. 1989. *Water resource management - integrated policies.* Paris: OECD.

UNICEF. 1993. *Barisal - A successful approach for sanitation.* Dhaka: UNICEF.

Water Supply and Sanitation Collaborative Council. 1994. 'Second Global Forum - Making the most of resources.' Meeting report of the second meeting of the Collaborative Council, Rabat, September 1993.

WHO. 1992. Follow-up survey on the 'Study of sale and performance of one slab-one ring latrines'. Users' attitudes after the one slab-one ring latrine pit is filled. Dhaka: World Health Organization.

WHO. 1993. 'New directions for hygiene and sanitation promotion.' The findings of a regional informal consultation New Delhi, 19-21 May 1993. New Delhi: World Health Organization.

WHO. 1993. Report on study on homemade (do-it-yourself) latrines - May-June 1993. Dhaka: World Health Organization.

Willingness to pay for/use water supply and sanitation

A.K.M. Nurul Islam, H. Kitawaki, Tokyo, Japan and M.H. Rahman, Dhaka, Bangladesh

THIS PAPER IS a sociocultural case study of Dhaka, the capital city of Bangladesh, on the willingness to pay for/use water supply and sanitation. 'Willingness to use' is a relatively new concept which can be defined as the maximum amount of desire one can willingly express for a certain commodity or service. Many technical, institutional, financial and legal factors determine the willingness to use. Affordability combined with willingness to use leads to willingness to pay. The low 'willingness to pay/use' is one of the reasons behind the unproportional success rate of much of the national and international efforts undertaken for safe water supply and sanitation. However safe water supply and sanitation is an urgent need in Bangladesh where the number of water-borne diseases cases per 100 000 persons is 51 000, much higher than other developing countries (WHO, 1992). One of the causes behind this alarming situation is low willingness to pay/use. Along with a brief description of the present situation in Dhaka city, efforts are taken to identify the problems behind the low willingness to pay for/use water supply and sanitation in light of the sociocultural aspects. This may provide useful information for other developing countries. Some improvement strategies are also proposed.

Present situation

Dhaka is the capital of Bangladesh. It is a mega city with a population of 6.54 million living in an area of 1345 sq. km. Per Capita GRP at current factor cost is TK. 7434 (Taka 40 = 1 US$), which is slightly higher than the national

GNP/capita which is TK. 7094 (Statistical yearbook, 1992). Like all other developing nations, variations in income of the people are quite high. The extremely poor people usually live in slum areas. About 40 per cent of the total city population are slum dwellers (Rahman & Islam, 1992). According to a report (Planning commission, 1991), the health conditions of urban slum settlements are extremely bad, and at any given time, 30 per cent to 46 per cent of this population suffer from diseases, most of which are related to environmental conditions — arising from pollution of water and air. Crude death rate for the urban slum area is 43.62 per thousand per year, which is four times the national average and six times the urban non-slum average. The 'under five' mortality rate is 150 to 180 per thousand, over 50 per cent higher than the national norm and almost double the urban non-slum rates.

The sole responsibility of providing potable water supply and waste water disposal service to Dhaka city lies with DWASA, Dhaka Water And Sewerage Authority. Apart from supplying the billed water, DWASA operates 1209 public stand posts for free water supply (DWASA, 1991). Most of these stand posts are located in areas that constitute a large portion of slum dwellers. It is calculated that each public stand post serves an average of 250 people. According to another study (CUS, 1988), about 50 per cent of the city's slum dwellers use water from DWASA service, 14 per cent of them get water from community shallow tube wells set up locally. The remaining 36 per cent do not have any access to safe water facilities. They obtain water from ponds or rivers. Therefore, this specific

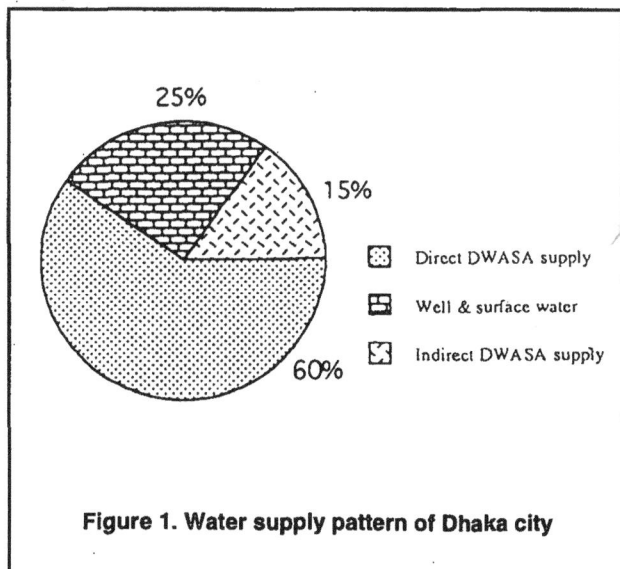

Figure 1. Water supply pattern of Dhaka city

25%
15%
60%

🔳 Direct DWASA supply
🔳 Well & surface water
🔳 Indirect DWASA supply

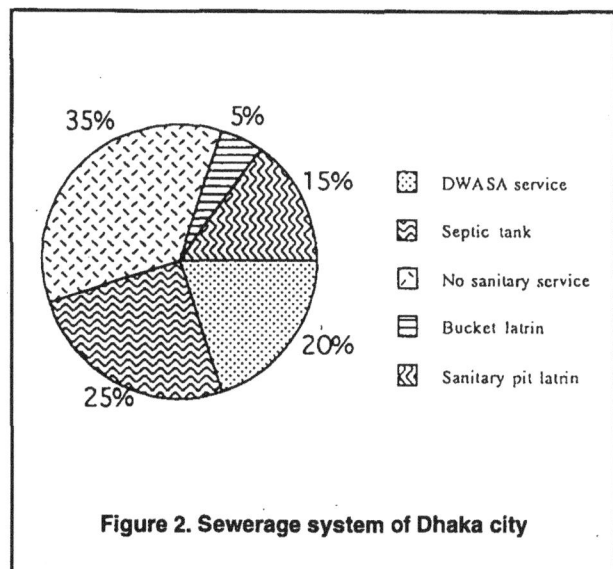

Figure 2. Sewerage system of Dhaka city

35%   5%
15%
20%
25%

🔳 DWASA service
🔳 Septic tank
🔳 No sanitary service
🔳 Bucket latrin
🔳 Sanitary pit latrin

44

36 per cent of the slum dwellers are the most susceptible group for health hazard. Another report shows (Islam, 1992) that about 25 per cent of the total city population are not served by DWASA. Most of this group use surface water for drinking purposes (fig. 1).

The situation in the sanitation sector is rather worse. Only 20 per cent of the city population have DWASA sewage service, and 35 per cent of Dhaka residents have no sanitation service at all (fig. 2). Those with no service use borehole latrines and must resort to open defecation, which deposits excreta directly into the local living environment. Dhaka City Corporation (DCC) now operates 20 public toilets that are extremely insufficient. Most slum areas have little or no sanitation services.

Causes behind the low willingness to pay/use

The city water supply and sanitation system suffers from various problems including those of technological, policy and planning and co-ordination and managerial aspects. Some of these can be explained by low willingness to pay/use, again which has its roots in the sociocultural background. Many cities of the developing world suffer from similar situations. To overcome these problems, some are discussed and identified herein.

Low service beneficiary

The DWASA sewerage system currently serves only 20 per cent of the Dhaka city population. This situation is not expected to improve over the next five years (Islam, 1992). As a result, people within the densely inhabited districts have been advised to build their own septic tanks to obtain safe sanitation. However, as there presently exists no design regulation, costs vary over a wide range. A typical 2000 litre septic tank required for a family of six persons may cost 200 to 300 US$, which is beyond the affordability of lower income people. Moreover, lack of construction land for the septic tanks also plays a key role. On the other hand, open defecation pollutes the surface, and sometimes, ground water. Most slum dwellers use surface water for washing, bathing, etc. and thereby become infected.

Improper sewage and water billing system

For residential metered connections, the water charge is TK. 3.49 per 1000 litres, however for non-meter connections, it is 22.62 per cent of the annual valuation of the holding (DWASA circulation, 1993). Currently, the cost for sewerage service is the same as the water bill charge. As the lower middle class people usually have a yard connection for their water supply, they are reluctant to receive the DWASA sanitation service because they would have to pay an additional service charge equal to that of their water bill. Ironically, most low income people have the more expensive non-metered connections since users must purchase meters. For lower middle income people,

the water and sewage bill may amount to 10 per cent of their monthly income, which is more than the World Bank recommendation of three per cent (World Bank, 1983). This reduces the willingness to use as well as pay. Moreover, the water charges are, at present, flat, regardless of water consumption and high water cost reduces willingness to use the water supply by lower income people.

Complicated and costly system for new connections

For a new connection, a potential sanitation service subscriber must spend a considerable amount of money to construct a pit. Then one faces many complicated procedures that require visiting and registering at various agencies such as DCC, DWASA (zonal office and service office), land taxation office and municipal taxation office. As middle or poor income people often live on limited land space, sometimes they need to construct their pit under a road, and for that purpose, they must obtain permission from the police and DCC. Furthermore, road reconstruction is the individual's responsibility. Experience has shown that incidental expenses are about 10 times higher than the official connection fee. These may be some of the factors behind low willingness to pay/use.

Low reliability of sewer system

In the city sewerage system, many manholes are damaged and sewers are blocked. Due to the flat terrain, 15 sewage lifting stations have been installed. However as pump station failure (mechanical or electrical), system overloading and sewage overflow remain common during the monsoon season, connection to the DWASA sewer system, which incurs a substantial expense, doesn't improve one's hygienic situation. Hence, people do not want to connect their toilet to sewer system.

Charge for public toilet

People must pay a certain amount to use public toilets. Although this rather low charge seems quite logical, it has became a factor for low willingness to use this service. A day labourer, for example, whose daily income may be as low as 50 Taka, will certainly be very reluctant to pay 0.50 Taka each time he or his family members use this service. For a family of six that uses the toilet a total of 10 times, one-tenth of the day's income will be incurred.

Location and number of public toilets and public stand posts

Since 30.31 per cent of the slums are illegal (CUS, 1988), the government can not technically support them by providing public stand posts or public toilets. The illegal slum dwellers are indirectly provided with this support since slum removal is not a short term priority. However, facilities are constructed a certain distance away from such slums, and dwellers must walk a long distance to reach these services that are available outside the illegal slums. Regrettably, this inaccessibility reduces their willingness to use. Moreover, the number of public toilets and public stand posts is quite insufficient. This is another important cause behind deficient beneficiaries in safe water supply and safe sanitation service. All these reasons reduce the willingness to use.

Operation period of public toilets and public stand posts

Although DCC has no time restriction on the operation period of public toilets, the operation contractors usually provide service from 6 a.m. to 10 p.m. only. As a result, people without private toilets are forced to defecate outside at night.

Water supply service is also not continuous. The supply period varies between 8 and 18 hours a day depending on location and season. Such inconvenience and drinking water pollution caused by low pressure during off service time is a cause for low willingness to pay/use.

Unapproved connection and under-reporting of bill

It is reported that 26 per cent of DWASA water is lost due to illegal connections and under reporting of water consumption (DWASA, 1984). The reason behind this 'unaccountable loss' lies in the users' rather high willingness to use and low willingness to pay.

Insufficient water pressure

The water supply network often suffers from low pressure; sometimes no pressure exists. This occurs due to an intermittent water supply. The water production (0.680 Million Cubic Metre per Day, MCMD) is much less than the estimated demand (1.260 MCMD) (DWASA, 1993). Moreover, 30 per cent line loss (DWASA, 1984) reduces the actual supplied quantity to 0.476 MCMD. As a result, second floor water access is denied. Hence, people have to install private underground storage tanks, overhead tanks and pumps. Despite having a willingness to use, this extra cost involvement reduces affordability.

Unauthorized practices

Some subscribers are carrying out unspecified private construction to extract water from the distribution system during periods of very low pressure, or when no pressure exists in the main. During such periods, the water main usually remains full or partially full of water. It is known that a handpump is connected to the water main to withdraw water. In another illegal manner, alternation of the service connection to draw water from the lower half of the pipe, instead of upper half, has been observed. These practices, to obtain water even when the main has a free water surface, cause non-availability of water to other subscribers, thus reducing willingness to pay/use.

Poor water quality

Contamination of sewage into the water supply system occurs as a result of low system pressure combined with pipe leakage. This problem emerges in two circumstances. The first mishap occurs in the narrow lanes where water supply mains are often placed inside sewer manholes due to insufficient road width, location of other service lines or other specific reasons. Constant exposure to corrosive sewer gases reduces pipe thickness. In situations where the water main is submerged in the sewer during periods when water supply pressure is less than external pressure, sewage may leak into the water main. The second contamination route is associated with the service connection. Portions of the lateral connection may lie in contaminated roadside water ditches. During periods of low or no pressure, contaminated water may enter through leaks originating from poor materials, poor workmanship or dilapidation. Poor water quality reduces the willingness to pay for/use water supply.

Discussion

Future strategies for hygienic improvement of the developing world's cities like Dhaka must be made with consideration of the overall picture of existing systems. The planned expansion of service should also be based upon future needs and affordability. Appropriate technology should be applied in other developing countries' cities as the strategies proposed here for Dhaka.

Low reliability, poor service, institutional obstacles, managerial problems and lack of awareness are among the compelling causes for low willingness to use. When this is combined with high costs for services, low willingness to pay is inevitable. This, in turn, results in poor service quality. This vicious circle should be broken with an affordable amount of money. For that, efforts should be taken to improve willingness to pay/use by some sociocultural initiatives. Some measures are proposed as follows.

Public awareness

A public awareness program for the promotion of hygiene should be started by providing up-to-date information on water quality and pollution. This should involve public media, NGOs, health service and women's organizations to reach each person in the community. Electronic media such as TV and radio would play a better role in this regard over other media, because about 75 per cent of the national population can not read. Although TV is a persuasive media which creates a sharp impact on public opinion, it is not accessible to all people. Radio is currently accessible for almost 100 per cent population. Hence, radio can play a vital role in the public awareness program. On the other hand, with the help of community activity, open defecation should be discouraged. Basic personal hygiene education should be included in primary schools, elderly persons' literacy and women literacy programs, public hospitals and health centres. Field level health workers should provide instruction on a house-to-house basis. This will increase the willingness to use.

Managerial aspects

A new tariff structure such as a progressive tariff system should be adopted so that people using more water would be charged at a higher rate. This is necessary, because people now use water for gardening and car washing at the same rate that is charged for drinking purposes. The concept of 'Some for all rather than more for few' would enable us to raise the willingness to use.

The sanitation billing system should be changed immediately. Instead of the present tariff system based on the holding valuation for non-metered subscribers, meters should be rented to provide equality among all users.

Service reliability

Good maintenance is a key point in maximizing capacity for waste water conveyance, thereby increasing willingness to use. Proper maintenance is also required for the water supply system. Essential facilities such as the pumping and lifting stations operated by DWASA should be protected against the adverse impact of floods. The unaccountable loss of 26 per cent must be eliminated for the betterment of the system. This may include public participation in reporting illegal activities. Identification of leaks and improvement of pressure in the existing system is urgently needed to eliminate the incidence of contamination during transmission and to reduce line losses. An immediate step could be lining the water supply pipes within manholes and through roadside ditches with a protective coating. Steps should be taken for increasing the water pressure along with increasing the production. Since line loss by leakage is proportional to supply pressure, projections to increase pressure through increased production may result in an increase in line loss. Therefore, increasing production alone may not increase deliverable supply pressure. Leak identification and repair is vital.

Institutional aspects

To increase the willingness to use, some institutional measures may be beneficial. These include, among others, simplification of formalities for new connections, provision of soft term loans and/or subsidization of pit latrine, septic tank or shallow well construction.

Community septic tanks

Community septic tanks with sewer pipes can be operated by DWASA for areas that cannot be readily covered by the sewerage system. This would be a stepwise improvement alternative that could be connected to future sewer lines.

References

1 CUS, 1988: 'Slums and squatters in Dhaka city,' The centre for urban studies., University of Dhaka.

2 DWASA, 1984 : Feasibility Study, Dhaka III project.

3 DWASA, 1991: Management information report, November, 1991.

4 DWASA, 1993 : Annual Report, 1992-93.

5 DWASA circulation, 1993 : appeared in Daily *Dhainik Bangla* on 93/11/16. [Bengali]

6 Islam, 1992 : 'Impact of Greater Dhaka Flood Protection Embankment on Domestic Waste Disposal and Environment', A report submitted by A.K.M. Nurul Islam to UNCRD, Nagoya, Japan, 1992.

7 Planning Commission, 1991: Report of the task force on social implication of urbanization.

8 Rahman, M.H., and Islam, A.K.M.N., 1992: 'Impact of wastes from slum areas on environment in greater Dhaka city', Proc. of the 8th International conference on solid waste management and secondary materials, Philadelphia, USA.

9 Statistical pocketbook of Bangladesh, 1992.

10 Statistical Yearbook of Bangladesh, 1992

11 WHO, 1992 : The international drinking water supply and sanitation decade, End of decade review, World Health Organization.

12 World Bank, 1983 : *Low cost sanitation : Instructors' guide*.

Developing private sector capacity

Derek Miles, WEDC, UK

THE AFFORDABILITY OF water supply and sanitation largely depends upon the choice of appropriate technology and its delivery through a network of efficient and cost-conscious providers. Uncertainties regarding long term funding, coupled with a growing appreciation of the inflexibility and declining productivity which can adversely affect the performance of large public sector organizations, have led to an increasing interest in reliance on private sector provision. However, many countries lack an experienced and resourceful private sector. So simple privatization is doomed to failure unless it is accompanied by a deliberate and well managed effort to select and train motivated entrepreneurs and ensure that they operate within a favourable and supportive business environment.

The paper explains the need for integrated technical and business training and draws upon the experience of the International Labour Organisation (ILO) in developing its Improve Your Business (IYB) and Improve Your Construction Business (IYCB) programmes. It also draws upon the principle of franchising to examine the scope for guaranteeing quality and providing training and other forms of support through an enabling institution.

The six S's

There are few countries which have not, reluctantly or otherwise, accepted the need for a greater reliance on private sector involvement in the provision of basic infrastructure. This is a trend which can be beneficial in increasing community involvement, generating local employment and securing improved efficiency and lower costs, providing that government retains a defined core role as promoter, facilitator and regulator so as to retain essential democratic control. The transition itself also needs careful management. In his paper to the 19th WEDC Conference, Duncan Morris[1] suggested three linked processes:

- decentralizing and delegating decision-making and local resource mobilization to the appropriate level of government but allocating financial support on the basis of national guidelines and criteria;
- making as much use as possible of the private sector (consultants and contractors) to prepare and implement public works projects, and also to manage, maintain and where necessary operate the assets created; and
- simplifying and streamlining the regulations and procedures relating to the provision and maintenance of basic public infrastructure.

This paper will argue that the success of the transition depends upon the application of six principles – the six S's:

- Subsidiarity
- Split responsibility
- System support
- Small enterprise focus
- Self help through developing business skills
- Sub-sectoral franchising

Subsidiarity

The principle of subsidiarity helps us to deal with the first of these issues, the appropriate level of government to which decision-making and local resource mobilization should be decentralized and delegated. In general, these actions should take place as close as possible to the level of the clients or ultimate beneficiaries. In the words of management writer Charles Handy[2] 'stealing people's responsibilities is wrong', a translation of the more formal definition of subsidiarity contained in a papal encyclical, Quadragesimo Anno, in 1941: 'It is an injustice, a grave evil and a disturbance of right order for a large and higher organization to arrogate to itself functions which can be performed efficiently by smaller and lower bodies...'.

Subsidiarity is easiest to achieve in what Duncan Morris describes as *community works*, defined as works 'initiated by clearly identifiable communities or common-interest groups (or even individual households) for the mutual benefit of their members'. The degree of effective subsidiarity in such circumstances depends upon mutual confidence, that is confidence by the client agency in the managerial capacity of local communities and confidence by those communities in the good intentions of the agency. The process of confidence building always takes time, and an ILO booklet *From Want to Work*[3] suggests that:

- the community itself must 'own' the project—not the authority or agency. *This means that the community takes all important decisions, while the role of the authority is to define the options to choose between, and give technical support;*
- agreements between community-based organizations (CBOs) and public authorities must be based on negotiation between equal parties. *This means that oral or (preferably) written contracts should be established spelling out rights and obligations of both sides, including the flow of funds, and the planned construction and maintenance arrangements. Public officials must refrain from being bossy and emphasize their role as civil servants.*

Split responsibility

As decision-making and responsibility for implementation is devolved to levels closer to local communities, the comprehensive role of government in delivering projects and services will have to be reallocated. Central governments' residual role is likely to be limited to the allocations of financial support on the basis of national guidelines and criteria, thereby setting the rules of the game and safeguarding legitimate national objectives. The remaining elements will include project identification and promotion, project design, implementation and operational management (including revenue generation, cost recovery and maintenance). An indication of the number of organizations and groups who can be involved in urban development is provided in Table 1.

Some work can be carried out by individuals and their families, and there may also be a role for community-based organizations and non-governmental organizations (NGOs). However *From Want to Work* cautions that:

> Some of these do not have great capacity or experience, and those which do easily become overburdened with work. Great care must therefore be exercised when commissioning community development work to an NGO to ensure that it is able to carry out its expected tasks.

System support

Table 1 emphasizes the complexity and variety of relationships that are involved in the infrastructure development process. Thus there is a case for establishing some form of intermediate 'system support' agency which will help communities to focus on practical objectives and will foster the emergence of capable and well motivated SSEs to provide the necessary products and services.

Table 1: Major and minor works: Actors
Source: ILO *From Want to Work*

Actors	Major Works	Minor Works
Individuals	• take paid employment • pay taxes	• contribute labour + cash • take paid employment • improve own house
Community	• execute sub-contract locally	• form development committee • decide priorities • collect local contribution • sign contracts • execute works
Small-scale contractors	• execute sub-contract	• specialist jobs
Large-scale contractors	• execute larger contracts • give out sub-contracts	• no role
Local government	• organize tendering • technical control • support contractor training	• technical support • issue contract to community • adapt building standards

---

Box 1. The AGETIP
(Agence d'exécution des travaux d'intérêt public contre le sous-emploi) in Senegal.

The AGETIP in Senegal was established in 1989 as a not-for-profit non-governmental organization with the following objectives:

- to create employment, particularly in urban areas;

- to provide vocational training, to improve the operational efficiency of the local construction industry and the effectiveness of public institutions;

- to demonstrate the scope for increased application of employment-intensive construction technologies, and

- to execute public works that are worthwhile in both economic and social terms.

It has been given a mission of 'owner's delegate' for a programme of small- and medium-size labour-based public works and services. AGETIP contracts out all engineer's duties for preparation and supervision to local consultants. Works and services are contracted out to artisans and small- and medium-size contractors. The agency carries out the whole management of the project, including inspection tasks. In 1990 AGETIP managed a program of about 130 contracts for works (average value US$ 80 000) and the same number of contracts for consulting services. Standard and computer aided procedures allow the agency to pay for works and services within a week.

Source: Lantran, J-M, and Egger, Phillipe

The World Bank and other influential donors/financing agencies have been attracted to channelling funds through autonomous agencies in recent years. The model for these agencies is the AGETIP (Agence d'exécution des travaux d'intérêt public contre le sous-emploi), which work largely with and through the private sector including preparation of bidding documents and inspection in the broad sense of 'owner's delegate' (in French 'mission de maitrise d'ouvrage deleguée').[4] A good example is the AGETIP in Senegal, which arose in 1989 in the wake of concerns related to the social effects of the structural adjustment programmes[5] (see Box 1).

The AGETIP approach depends for its success on bypassing cumbersome and bureaucratic government procedures, paying competitive salaries to a comparatively small number of well motivated national staff of high calibre and making extensive use of the private sector so as to achieve a greater degree of 'responsiveness', 'flexibility' and 'efficiency' than conventional government structures. These agencies are mostly relatively new, and their performance in the longer term is yet to be tested. They may be tempted to emphasize 'efficiency' at the cost of 'responsiveness', unless they bear in mind Duncan Morris's advice that 'the beneficiaries (should) become the clients, with donors and governments placing more reliance on their innate common sense than they have in the past'.

AGETIP-type agencies have been replicated in a number of countries (notably Benin, Burkina Faso, Mali, Mauretania and Niger), and are likely to continue to demonstrate success while they benefit from a significant flow of external resources. It is, however, difficult for a single organization to combine development activities with project execution imperatives, and conflicting pressures may emerge. If the objective is to enable firms to become self-reliant, it may be better to concentrate on training, enterprise development and fostering a corporate approach among the contractors themselves. These were the priorities mentioned by Kirmani and Blaxall in their 1988 World Bank Discussion Paper[6], where they proposed the following areas for action in World Bank-financed projects:

(a) improving contracting and contract administration policies and practices;

(b) improving the business environment of the industry;

(c) improving the efficiency of contractors;

(d) developing the institutions of the construction industry; and

(e) research and development related to the construction industry.

Small enterprise focus

Small-scale enterprises (SSE) offer a variety of advantages by[7]:

- making use of materials and resources that may otherwise not be drawn into the development process;
- creating jobs at relatively low capital cost (small-scale contractors are more liable to choose employment-intensive solutions than large contractors);
- providing a vehicle for introducing a more equitable income distribution;
- employing workers with limited formal training, who then learn skills on the job and provide a pool of local skills that will favour future economic development;
- improving forward and backward linkages between economically, socially and geographically diverse sectors of the economy;
- providing opportunities for developing and adapting appropriate technological and managerial approaches;
- promoting special subcontracting arrangements and acting as ancillaries to large-scale enterprises; and (last but far from least in a turbulent environment)
- adapting flexibly to market changes.

Self-help through developing business skills

In developing countries most small contractors are practical people, who may have experienced trade or vocational training but have rarely been introduced to basic management and business training in a form that is relevant to their practical needs. The result is that their businesses are fragile and frequently fail, resulting in a loss of scarce national resources and embarrassment for their clients. The construction industry development problem is thus predominantly a management development problem, since most of the difficulties which beset the industry could be overcome if managers within it could engender greater trust in their skills among clients and resource providers.

In its search for a means to assist this target group, the ILO Construction Management Programme drew upon the experience of another ILO programme - IYB or 'Improve Your Business'. IYB was based on a Swedish training package 'Look after your firm', which consisted of self-teaching material that introduces the user to basic management techniques such as business analysis, financial analysis (accounting, key ratios and so on) and activity and financial planning. The material is packaged neatly in the form of a 'handbook' and a companion 'workbook', and includes checklists and a reference guide, together with advice on how to prepare an action plan and practical suggestions on how to bring the plan to fruition. Thus the obvious solution was to develop a system based on the IYB approach but taking into account the special operating conditions in the construction sector. An equally obvious title for the system was IYCB or 'Improve Your Construction Business'.

For small contractors, estimating and tendering are crucial activities. Construction is a fiercely competitive industry; profit margins are often low and a small mistake on pricing a tender document can make all the difference between a worthwhile profit and a serious loss. Furthermore each individual project is taken on at a fixed price and represents a significant proportion of the contractor's annual turnover, so one serious error in pricing a single project can undermine the stability of the enterprise as a whole. Yet many small contractors lack even the most elementary grasp of cost accounting, and estimates could frequently be more accurately described as 'guesstimates'. Thus the first IYCB handbook and workbook deal with the topic of *Pricing and Bidding*[8]. The two priority topics of project planning and productivity are essentially site activities, so the second handbook and workbook cover *Site Management*.[9] This leaves the range of activities concerned with managing the enterprise as a commercial entity, including ensuring a reasonable balance between workload and resources, which comes under the general heading of *Business Management*.[10] These six business-related books form a basic framework for small contractor training, permitting the trainer to concentrate on the preparation of local case studies and exercises.

Sub-sectoral franchising

Although the IYCB package is suitable in its present format for building and general contractors, there is scope for supplementing it with complementary technical training packages directed at specialist sub-sectors which present potential market opportunities for construction-related small-scale enterprises. These opportunities can

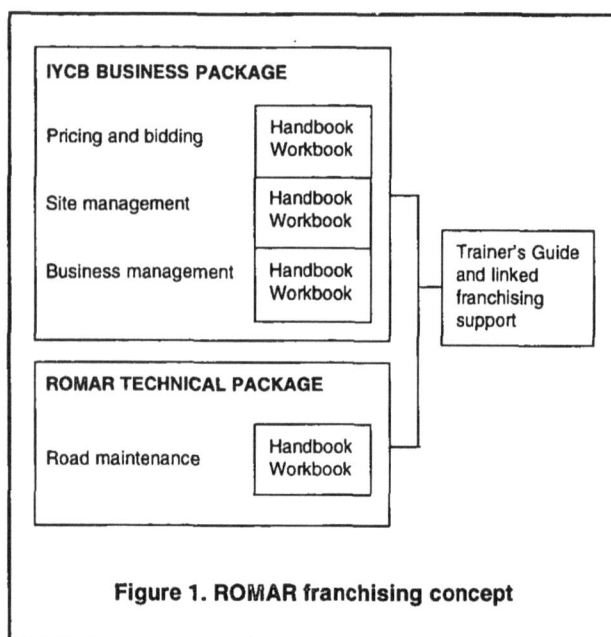

IYCB BUSINESS PACKAGE

Pricing and bidding	Handbook Workbook
Site management	Handbook Workbook
Business management	Handbook Workbook

Trainer's Guide and linked franchising support

ROMAR TECHNICAL PACKAGE

| Road maintenance | Handbook Workbook |

Figure 1. ROMAR franchising concept

occur either through new developments or through the growing emphasis on involving the private sector in activities that were previously the exclusive preserve of direct labour or force account, such as road or building maintenance (or many aspects of work related to water supply and sanitation). A project to apply the concept to the development of labour-based road maintenance contractors is now underway in Lesotho, and this will lead to the development of a complementary technical training package consisting of a *road maintenance and regravelling (ROMAR) handbook* and *workbook*. This approach could readily be extended into other specialist areas, including water supply and sanitation. Together the general and specific handbooks/workbooks could be combined with an outreach consultancy and advisory service to provide the basis for a franchising approach to sub-sectoral small business development (Figure 1).

The scope for replication

Many countries appreciate the potential for achieving savings through greater involvement of private enterprises in infrastructure construction and maintenance. Unfortunately too many interventions in this field have been piecemeal and unco-ordinated, with the result that resources have repeatedly been spent on duplicating previous experience rather than advancing the state of the art. It is therefore worth emphasizing that the need for these activities is going to persist and that there is real scope for achieving economies and improved results through the widespread application of a system such as IYCB, since:

o the target group of small contractors is numerous in most countries, and is growing rapidly with the impact of privatization in those countries which used to emphasize central planning and control;

o the management problems faced by the target group do not differ significantly from country to country;

o the IYCB material has benefited from the earlier experience of its two parent programmes, and seems to be neither too complex nor too simplistic for the needs of the target group;

o many of the problems faced by contractors stem from a lack of commercial and financial knowledge and experience, so the emphasis of the IYCB material on these aspects is very relevant.

In all cases IYCB projects will continue to feature a strong institution building dimension, and will seek to tackle the policy constraints resulting from an inadequate regulatory and contractual framework as well as providing direct assistance in the form of training, coaching and consultancy.

References

1 Morris, Duncan, 'Thinking Things Through' in *Water, Sanitation, Environment and Development: Proceedings of the 19th WEDC Conference*, WEDC, Loughborough, 1993.

2 Handy, Charles, *The Empty Raincoat: Making sense of the future*, Hutchinson, London, 1994.

3 ILO, *From Want to Work: Job creation for the urban poor*, ILO, Geneva, 1993.

4 Lantran, J. M., *Contracts for Road Maintenance Works Agreements for Works by Direct Labor*, World Bank, Washington, 1991.

5 Egger, Phillipe, *Travaux publics et emploi pour les jeunes travailleurs dans une économie sous ajustement: L'expérience de l'AGETIP au Sénégal*, ILO Interdepartmental Project on Structural Adjustment: Occasional Paper 2, ILO, Geneva, 1992.

6 Kirmani, S. and Blaxall, J., *The Construction Industry in Development: A strategy for Bank Assistance*, World Bank Policy and Planning Research Staff Discussion Paper, Washington, 1988.

7 Neck, Philip A. and Nelson, Robert E., *Small Enterprise Development: Policies and programmes*, ILO, Geneva, 1987.

8 Andersson, Claes-Axel; Miles, Derek; Neale, Richard and Ward, John; *Improve Your Construction Business Handbook 1 and Workbook 1: Pricing and Bidding*, ILO, Geneva, 1994.

9 Andersson, Claes-Axel; Miles, Derek; Neale, Richard and Ward, John; *Improve Your Construction Business Handbook 2 and Workbook 2: Site Management*, ILO, Geneva, 1994.

10 Andersson, Claes-Axel; Miles, Derek; Neale, Richard and Ward, John; *Improve Your Construction Business Handbook 3 and Workbook 3: Business Management*, ILO, Geneva, 1994.

Private sector involvement in water services

J. M. Moss and J. P. Terme, Lyonnaise des Eaux, France

IT IS BECOMING more frequent for the private sector to be involved in the provision of public services such as water and waste water.

There is a range of ways in which a dedicated services company from the private sector can be involved in the management and provision of municipal services. These range from a very restricted provision of technical services, through a variety of contracts involving increasing responsibility to full ownership of the supply system.

This is illustrated in the chart attached which indicates the main structural alternatives in a hierarchical manner. The principal stages of this are as follows:

(1) The division between the provision of the services through a contractual relationship with a municipality or a State and the situation where the private sector is the outright owner of the supply company and assets.

(2) Within the contractual subdivision there is a division between the provision of services only, the provision of services with working capital and the provision of all investment and operation.

Starting with the simplest arrangement which involves the least level of private sector responsibility and progressing to total private sector ownership the main alternatives are characterized as follows :

Ancillary operations

The operating contractor provides a well defined technical or management service in support of the municipality's management resources. For example a municipality wishes to implement a mains rehabilitation exercise. Pressures on its resources are such that it finds that it is beneficial for it to engage an operational contractor who can undertake the management of the programming, works implementation and control, consumer liaison etc. The municipality continues to operate the network using its own staff while the contractor completes the specialist work.

Operation and maintenance (O&M)

The operating contractor carries out the day-to-day operation and routine maintenance of a specific geographical sector or defined level of responsibility.

In this situation the operator would be paid for the work done in accordance with the contract price structure.

Maintenance expenses would be reimbursed against schedules of rates and where appropriate priced bills.

Operation (with working capital only) (Affermage)

The operator is contracted to carry out all routine operation and maintenance. The operator is not involved in the financing of capital works. He is only called upon to provide the working capital required for the actual operation. It is likely that he will be required to provide a consultancy service in respect of any new works which may be proposed. The operator may be involved in the operation right through to customer billing, or he may be paid by the undertaking who bills the customers.

Operation with new capital investment (Concession and BOT)

These forms are similar. The operator finances all costs for the installation of the utility (new works and renewals) and the working capital required for its operation and maintenance. In the most highly developed contracts (Concession) the operator is responsible for billing the individual customers. The price of the water sold is controlled under the contract and also includes a proportion which is collected by the operator on behalf of the undertaking. In BOTs, which are slightly more simple, the customers are usually charged by the undertaking and the operator is paid in accordance with rates set in the contract. In either case, these payments can include productivity bonuses, or shares in increased profits if appropriate. At the end of the contract period the installations are handed over to the municipality unless the contract is renewed or extended.

Joint venture

In this case the private operator and the municipality create an appropriate special purpose company in which they both have a shareholding. This joint venture company undertakes the provision of the full service and charges the customers direct. The joint venture can either operate under a contract with the municipality or under a licence arrangement. Remuneration to the shareholders is through the payment of dividends. Often the private sector partner also provides special management and technical support.

Private ownership

In this case the operating company and the infrastructure and assets are owned entirely by the private sector. The regulation of the private monopoly thus created is normally by some form of statutory control or licence agreement.

Whilst from the foregoing it may appear that each category is clear cut, in reality this is not the case as there is a great deal of flexibility. Eventually the solution is

adapted to meet the needs of the individual case in question.

This can be illustrated with examples from China, Indonesia, Malaysia, Viet Nam, Australia, Argentina, North America and Europe. Each case shows how the private sector service company focuses on different customer needs and political requirements. They also show how the solutions are adapted to the cultural, economic, legal and financial structures of the countries in question.

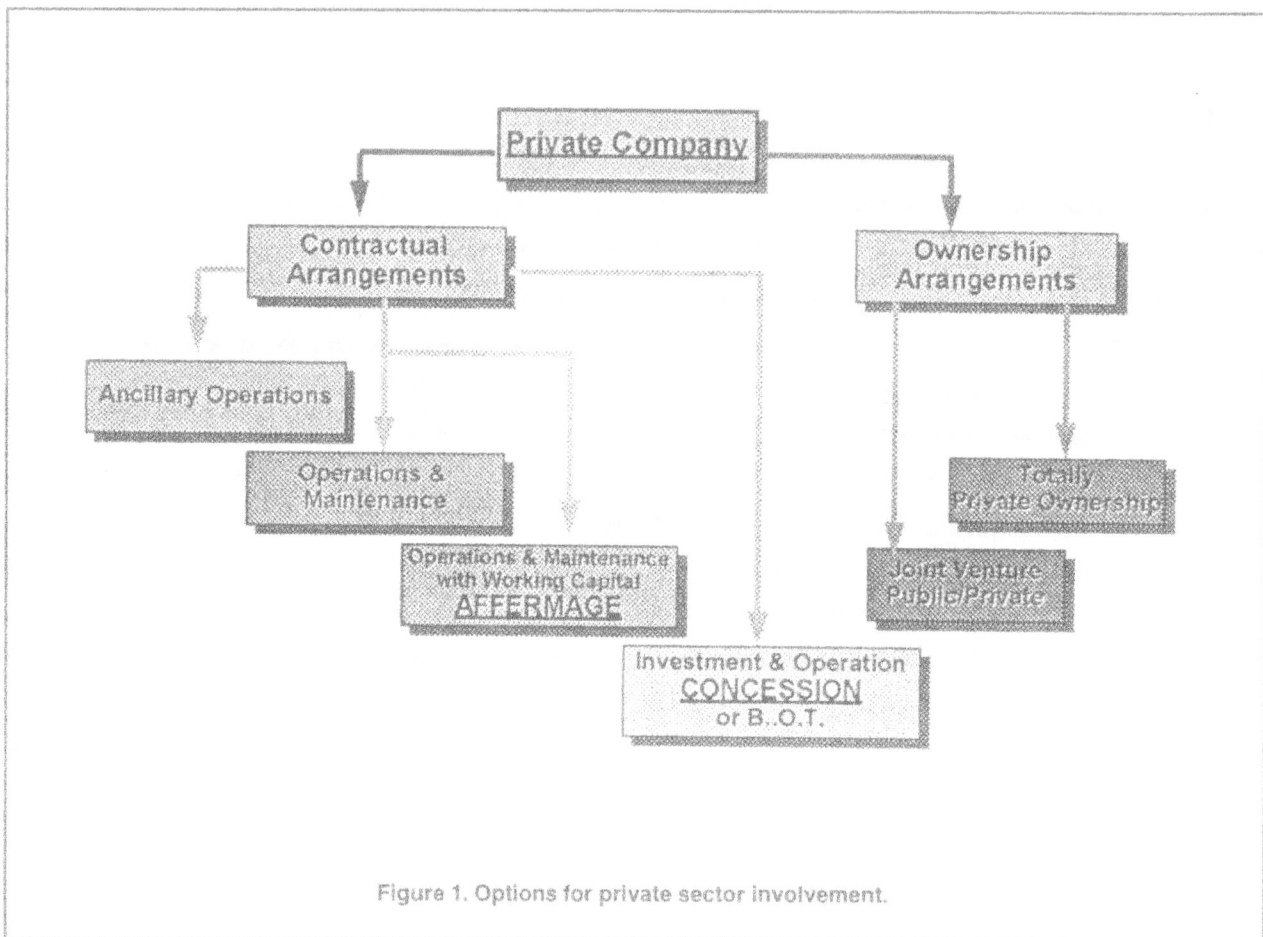

Figure 1. Options for private sector involvement.

Institutional strengthening for rural water supplies

Kevin Sansom, ODA, India

THIS PAPER OUTLINES an assessment methodology for developing an institutional strengthening programme for the management of the operation and maintenance of regional rural piped water supplies schemes (RRPWSS). This method has been used to do an institutional analysis of organizations responsible for regional schemes in rural areas of north-west Maharashtra in India. Some of these RRPWSS cover large areas serving up to 50 or 60 villages. The institutional strengthening proposals for the participating organisations in NW Maharashtra, are discussed as well as experiences to date in evolving, planning and implementing these and other proposals.

Regional rural piped water supply schemes in north west Maharashtra

In the drought prone areas of NW Maharashtra, village drinking water supplies with ground water sources have become unreliable in some areas, due to over-exploitation by farmers irrigating cash crops, plus the general population growth in the area. A programme of RRPWSS is under way to serve problem villages. The local zilla parishads (district councils) have been made responsible for the operation and maintenance (O&M) of these regional schemes, some of which involve expensive water treatment and high lift pumping. The increase in responsibility for the zilla parishads (ZPs) and the villages is part of the government's decentralization policies.

Jalgaon ZP, for example, will be responsible for O&M of 20 RRPWSS serving 216 villages. At present it has taken over O&M of 10 of the smaller RRPWSS. Eight of the remaining schemes which are soon to be handed over, are currently managed by the Maharashtra Water Supply and Sewerage Board (MWSSB), who designed and supervised the construction of these schemes, as well as two ODA supported schemes currently under construction, in Jalgaon District, which will serve approximately 130 villages. The Overseas Development Administration (ODA) financed project has a number of components including construction, community development, health education and preparation for O&M. The zilla parishads are responsible for the RRPWSS from the source up to the elevated service reservoirs in each village. Individual village water committees are responsible for the O&M of the piped water distribution system within their own village and water tax collection.

It is necessary to examine the capacity of the zilla parishads to take on these new responsibilities and de-velop, through an adaptive process approach, an institutional strengthening programme to ensure sustainable and reliable water supplies.

Identification of institutional strengthening (IS) proposals

In order to identify suitable IS proposals that take account of the complexities of the systems within which the potential IS programmes will develop, a methodology for the identification of IS programmes for institutions engaged in the management of RRPWSS is proposed. The methodology is intended to be comprehensive but not be too demanding on resources. To endeavour to achieve a holistic view, it is proposed that the existing situation and the potential for change be examined from a number of different perspectives including:

(a) **Background**—Existing and proposed future demands to be placed on the relevant institutions or departments.

(b) **The broader perspective** — Examine the physical, socioeconomic and political environment in which the relevant institutions are functioning. Using the Systems Approach, identify key decision-makers and processes. Undertake a Stake Holder Analysis to determine to what extent the key organizations have vested interest in the success of the project/programme.

(c) **The local perspective** — Case studies of typical systems that are currently being managed by the relevant institutions, such as regional piped water supply schemes, examining problems that emerge. Consideration should be given to what, why, how and when important factors and events occur and who is responsible. These studies should be prepared on the basis of site visits and consultations. There should be a focus on the level of service provided to the users. System performance indicators are helpful in this regard.

(d) **The economic perspective** — Cost recovery and sustainability should be examined, both in the short and long term, under this heading. This would involve an examination of predicted costs, tariff design options, cost recovery strategies, and measures needed to achieve sustainability, including community development. An assessment of the overall or national economic environment in which the systems are being managed, can be useful in indicating the scale of the problems to be dealt with and the likelihood of success.

(e) **The organizational perspective** — An assessment of the capabilities of the relevant institutions to meet existing and proposed future demands placed on them, focusing on the area of interest, for example, water supply. The following key areas which are critical to an institution's success should be examined:

- organizational autonomy
- leadership
- management and administration
- consumer and commercial orientation
- technical capability
- developing and maintaining staff
- organizational culture
- interactions with key institutions

(Wash technical report no 37, 1988)

(f) **Options for change** — Different management options, perhaps involving other institutions and the private sector should be considered. Issues such as cost effectiveness, level of service, meeting of basic human needs, community participation, regulatory frameworks, ownership of assets and accountability need to be borne in mind. Options for potential institutional strengthening programmes should then be examined.

By viewing and analysing a complex situation from a number of different perspectives in the manner suggested above, a better understanding should result, in rather the same way that a variety of views of a complicated object can lead to more comprehensive understanding. A clear appreciation of the analysis of the problems and the resultant proposals is required, particularly by the organizations that need strengthening, because it should be these institutions who plan and implement any institutional strengthening programmes.

The above categories (a to f) are intended to form the main sections of an identification report and have been used in the assessment of the management of O&M of RRPWSS in NW Maharashtra. The following section is a summary of that assessment.

The need for institutional strengthening in NW Maharashtra

The lack of capacity in the zilla parishads to undertake the management of the O&M of the RRPWSS and provide reliable supplies, stems from a lack of financial and human resources. These are problems with the reliability in terms of quality, quantity and distribution on some existing RRPWSS. On those schemes currently managed by Jalgaon ZP, the engineers responsible for management are seconded from the irrigation department. Despite being hard working, they have a diverse range of responsibilities and are, therefore unable to concentrate on O&M. Technical O&M staff in the field are paid very low wages, about a third of what equivalent MWSSB staff receive.

Staff numbers are less than half of what was recommended by the MWSSB. Limited 'on the job training' is the only training provided to date, although there are plans for O&M training as part of the ODA project.

The lack of resources is due mainly to poor tax recovery and low water taxes. There are plan to increase tax levels. Persuading the people to pay the much higher taxes is a big challenge, involving the provision of reliable supplies and community development work. Subsidies may be required at least in the short term.

The constraints to effective management and provision of reliable supplies are also institutional in nature including the following:

- There is a multitude of departments, organizations and committees responsible for drinking water supplies in rural areas. There is, therefore, a lack of accountability and clear lines of responsibility.
- The hierarchical nature of local government inhibits adequate delegation of duties, as well as vertical and horizontal communication.
- The decision-making processes for important aspects such as investigating inadequate supplies and planning scheme augmentation are reactive and rather long, involving many organizations. A more proactive streamlined approach would assist in being more responsive to problems and securing reliable supplies.
- The commercial and consumer orientation of the water supply institutions could be improved. This is particularly true in terms of their relationship with villages who are responsible for the management of the piped distribution systems within the village. A more contractual relationship would be preferable, to ensure both parties benefit, so that greater commitment should result.
- The limited organizational autonomy of the zilla parishads to agree their own objectives and policy, as well as plan and implement is a constraining factor. Considerable co-operation between the state and district level organizations will, therefore, be necessary.

Many of these constraints are common to other states and countries. The MWSSB generally have good technical and managerial capability, although improvements could be made in monitoring against performance indicators, management information, and community participation. It also has limited organizational autonomy from the state government.

Institutional strengthening proposals

On the basis of the writer's investigations and consultations, the institutional strengthening proposals for NW Maharashtra are summarized below under five main headings. Experience has shown the need to address institutional problems through system wide efforts as part of an integrated design, it is important, therefore that

improvements be made in each of the following five areas:
(Wash technical report no 49, 1986)

Structural and organizational adjustments

A separate water supply division should be created in the ZP to develop expertise and ensure managers are not distracted with other tasks. This will also improve accountability. Experienced staff should be seconded or transferred from the MWSSB.

A co-operation agreement between the ZP and the MWSSB is proposed to establish clearly how the two organizations should co-operate on such matters as: training, monitoring, data collection, planning scheme augmentation, provision of advice, conducting studies, transfer of staff etc. This is proposed as an interim arrangement until an integrated river basin authority or a rural water supply authority is established.

Management development

A more systematic use of the following ideas should be encouraged.

- management by objectives, perhaps using techniques such as the logical framework;
- development of strategies to overcome persistent problems;
- adoption of the cyclic process of — evaluate current practices — set objectives — define priorities — develop a detailed plan in terms of programme, resources and costs — check plan against objectives — implement plan — monitor plan and systems being managed — provide appropriate feedback to staff and senior management — revise plans and priorities;
- development of comprehensive handover agreement with villages;
- delegation of duties, preferably mutually agreed with the subordinate and in writing;
- a flexible approach to village problems and a preparedness to consult all sections of the community.

Systems and procedures

Effective management of water supply institutions requires information on whether or not an organization is meeting its objectives, on how efficiently an organization's resources are being used to meet these objectives, and the basis for taking timely corrective action when objectives are not being met effectively. Systems and procedures should be designed so this information can be obtained by the simplest means.

A maintenance management system is proposed, which should include: development of an annual O&M plan, preventative maintenance schedules, work order systems, budgets based on recent experience, performance and cost monitoring, equipment history files, stores and materials inventory files.

Resources

A recruitment and procurement plan should be developed once detailed plans and resource requirements have been agreed. The amount of funds available will of course depend on the expected cost recovery and subsidies. A detailed estimated cash flow programme over the coming years will assist in balancing expenditure with income, particularly where cost recovery and commitments are changing.

Training systems development and skill training

An extensive training programme has begun as part of the ODA project in Jalgaon, Nasik and Dhule Districts, focusing on community development, health education and technical training. A more detailed focus on effective management, motivation theory, leadership, teamwork, delegation, monitoring and feedback, management by objectives, budgeting and maintenance management systems, would be beneficial. Technical training should focus on the application of skills as well as their acquisition.

Institutional strengthening process

As part of the current integrated project in Maharashtra supported by ODA, workshops have been regularly convened to consider problems, co-ordinate and develop a shared vision among the participating organizations. Discussion papers have been produced to raise anticipated problems and possible solutions. It is important, however, that the preparation of plans and the implementation and monitoring of the plans is done by the local institutions themselves, if we are to achieve commitment and sustainability. Regular meetings are held to co-ordinate the various components of the project and deal with problems as they emerge. The recruitment of additional staff as well as community development and management consultants is necessary to provide advice and support. In general terms a learning adaptive approach to institutional strengthening is used.

Progress

Detailed district O&M management plans are being developed in NW Maharashtra to tackle the difficult task of providing reliable water supplies in the regional schemes. Some additional staff and consultants have been or are in the process of being recruited to provide assistance. Many staff working for the various organizations involved with the project, show a keen interest to participate and contribute, this is despite the many constraining factors to institutional strengthening, some of which are listed below:

- Key staff in participating organizations generally have many other work commitments outside the project, and transfer of key project staff sometimes occurs as part of the usual government practices.

- An important state government policy document which sets out policy on water taxes, staffing arrangements and the regulatory framework had not been released at the time of writing this paper. The planning of some aspects of O&M and scheme handover has been awaiting this new legislation.
- The local government system of checks and balances that was originally introduced during the colonial period, tends to act against effective management of services and permitting the change necessary for institutional strengthening.
- The regional water supply schemes in this region that cover large areas serving many villages, are generally costly to operate and maintain due to expensive treatment requirements, high lift pumping heads, long leading mains etc. To achieve full O&M cost recovery relatively high water taxes are required. The willingness of people to pay these taxes has yet to be firmly established. There are, therefore, uncertainties over cost recovery which leads to a reluctance to plan for the high expenditure necessary to provide reliable supplies.

Conclusions

When identifying a potential institutional strengthening programme, it is worthwhile to undertake comprehensive investigations examining the situation from a variety of different perspectives as has been discussed, in order to build an understanding of the complexity of the system in which the relevant institutions function. The initial design of the programme can then be done with a better awareness of the scale of the problems and risks involved.

The actual strategies and plans that are developed and implemented should be done by the local institutions themselves, with outside consultants or advisors facilitating the process and providing guidance and suggestions as the need arises. Improving the capacity of the local water supply institutions has been deliberately referred to in this paper as 'Institutional Strengthening' rather than the usual 'Institutional Development'. Development gives the impression of outsiders coming in to develop as if they were considering the development of a piece of land. Institutional strengthening is a better term because it implies building from within — which is what is required for commitment to sustainability.

The zilla parishads and water board in Maharashtra both have limited organizational autonomy. Experience elsewhere, notably studies undertaken by the WASH organization, have shown that lack of organizational autonomy is a key constraint to institutional strengthening. It is true that the achievement of results from an IS programme are likely to be considerably slower when an organization has limited autonomy. But the potential for replicability and benefiting from lessons learned is greater with a series of inter-connecting organizations such as those in local government in Maharashtra. Development projects and programmes should be regarded as input into the longer term development process of 10, 20 years or more.

The affordability of the water taxes on the rather expensive regional rural water supply schemes in Maharashtra cannot be fully known until reliable cost effective water supplies are provided and community development work has taken place through strengthening of the local institutions. Cross subsidies from the urban sector may need to be considered in the future, and/or the use of more local technical water supply solutions.

References

[1] Cullivan et al, WASH Technical Report no 37, *Guidelines for Institutional Assessment of Water and Wastewater Institutions*, Arlington, USA, 1988.

[2] Edwards, D.B., WASH Technical Report no 49, *Managing Institutional Development Projects: Water & Sanitation Sector*, Arlington, USA, 1986.

[3] Sansom, K.R., *Management of O&M of Rural Piped Water Supplies in NW Maharashtra*, UMIST, UK, 1994.

[4] WASH Technical Report no 63, *Guidelines for maintenance management in Water Supply and Sanitation Utilities in Development Countries*, Arlington, USA, 1989.

SECTION 3

ENVIRONMENT AND SANITATION

Urban drainage - the alternative approach

Dr. R.Y.G. Andoh, Hydro Research and Development Limited, Avon, UK

THE PAPER PRESENTS an alternative drainage philosophy and strategy which mimics nature's way by slowing down (attenuating) the movement of urban runoff. This approach results in cost-effective, affordable and sustainable drainage schemes. The alternative strategy can be described as one of *prevention rather than cure* by effecting controls closer to source rather than the traditional approach which results in the transfer of problems downstream, compounding the problem, resulting in its cumulation and, the need for large-scale centralized control.

The alternative strategy is set in context relating to the evolution of current practice and the implications of fragmented institutional responsibilities. Issues relating to socio-economic factors and appropriate development are discussed and the alternative strategy is shown to incorporate tenets of a holistic approach.

The paper concludes by recommending the adoption of the alternative strategy for the provision of urban drainage infrastructure in developing countries. It suggests that this *paradigm shift* should help developing countries to leap frog the developmental stages in their provision of effective urban drainage infrastructure for their rapidly burgeoning urban centres.

Background

The current world population of over five billion people is estimated to increase to over six billion by the end of the millennium. Of the estimated 90 million people currently added to the global population, each year, 94 percent are in developing countries. The poor of the developing countries are moving into urban areas and as a result the urban centres are the fastest growing areas of these countries. The majority of the increased global population will therefore live in the burgeoning urban centres of the developing countries. This will no doubt place un-parralled demand on the already inadequate urban drainage infrastructure in most developing countries.

The resolution of problems associated with infrastructural provision in most developing countries currently follows along the traditions of the developed countries. Often, this is not appropriate for the locality (Sonuga, 1993). A review of urban drainage practice shows that, in the past, the philosophy has been based on conveying peak flows of municipal waste water and storm runoff away from the urban areas as quickly as possible. This has resulted in downstream flooding and heavy pollution of receiving waters.

The problems of urbanization manifest and currently being dealt with in both developed and developing countries such as flooding and pollution of ecologically sensitive urban streams, will no doubt grow worse in the developing countries. If the cost estimates of £107 billion (1992 price base) attached to the meeting of the European Council (EC) urban wastewater treatment directive by the EC member countries (Wright, 1992) is anything to go by, then it is clear that the provision of urban drainage infrastructure along the conventional approach is going to be unaffordable for the developing nations.

This paper presents an alternative drainage philosophy and strategy, based on the philosophy of the single pipe system (Smisson, 1980), which aims to mimic nature's way by slowing the movement of storm water from urban areas, encouraging the infiltration of relatively uncontaminated rainfall runoff to help maintain base flows in rivers and the beneficial re-use of rainwater through distributed storage close to source.

Case studies are presented demonstrating the significant cost savings that can be realized by adopting the alternative approach.

Evolution of conventional urban drainage systems

The urbanization process has involved the growth of communities with people living closer and the conversion of open ground, that absorbed rainwater, to impermeable pavements and buildings. Associated with this has been the accompanying reduction in areas where the resulting storm water, following rainfall, could be absorbed into the soil. This resulted in flooding in the vicinity of households and as a result, open channels were constructed to convey runoff from roads and roofs, away from properties to prevent flooding.

The increasing population concentrations associated with the urbanization process also resulted in increases in waste generation. Household wastewater were connected (disposed off in the nearest open channel) and in turn created problems of smell. As a result, the open channels constructed to alleviate flooding were covered creating combined sewer systems.

Sewage treatment (initially via sewage farms) evolved from the need to alleviate the problem of pollution resulting from discharges from combined sewers into receiving waters close to the centres of population. For example, the Thames River through London in the UK, was becoming foul smelling between 1862 and 1864. Sewers were therefore built down to the estuary below London to take London's wastewater and discharge it straight into the estuary. A number of years later, however, problems were again becoming apparent so sewage treatment started in the late 1800s.

A review of the pattern of development of urban drainage described, shows that mankind's response has characteristically been one of finding a cure to an observed problem. Mankind has traditionally operated and still to an extent operates under a feedback law. A control action is sought and implemented only when an undesirable state of affairs is observed.

A publication titled 'Urban Drainage The Natural Way' (HRD, 1992) which summarizes the proceedings of a two-day conference held in Oxford in the UK in 1992 (Conflo '92) considers how 'Source Control' a collective term used to describe the management of runoff at or near the point of impact of rainfall and before it reaches the traditional piped drainage and sewer systems of urban areas, might be used to mitigate the impacts on the natural water environment, of storm runoff from urban development. Institutional arrangements are highlighted as not being wholly conducive to the use of the Source Control methodologies for the implementation of engineering solutions to problems associated with urban drainage.

Institutional issues

In most countries, the institutional arrangements pertaining to responsibilities and control of the various stages (facets) of the water cycle have typically been fragmented with, for example, one institution responsible for municipal water supply, wastewater sewerage and treatment; another responsible for land drainage and urban runoff drainage systems; and yet another for the drainage of highways and urban roads. This fragmentation has contributed to unequal attention and unequitable allocation of resources to the various facets of water resources management and has not been conducive to the implementation of an integrated watershed or catchment approach to urban drainage owing to the imposition of artificial boundaries. Cross-connections, wrong connections and combined sewer overflows mean in effect that the traditional descriptions of foul, combined or surface water sewerage systems of urban drainage, which have typically formed the basis for the divisions in institutional responsibilities, are not strictly correct. In the urban environment, the interactions between the various wastewater networks (e.g. combined sewers, highway drainage and land and surface water drainage systems) means in effect that wastewaters derived from sources under the jurisprudence of one institution could in effect ultimately be disposed off into a receiving water through a network in the jurisprudence of another institution.

If the overall interests with regards to mankind's interactions with the water cycle in the urban environment is stated as one of; 'the provision of adequate quantities of safe (potable) drinking water supplies, the safe disposal of all urban wastewaters, the maintenance of water re-

sources and the prevention of adverse aquatic environment impacts', then it is suggested that the division of institutional responsibilities be along the lines of 'service provider' and 'regulator'.

The service provider in this context would be an integrated institution with overall responsibility for the abstraction and supply of potable water supplies and the safe collection and disposal of all wastewater sources from the urban environment including stormwater runoff from roofs, road and other impermeable surfaces. Boundaries for such institutions would not be along the lines of administrative regions but rather on receiving water catchment or natural watershed basis.

The regulator would then have the function similar to that of environmental protection boards and agencies (essentially a policing function/role) with regards to the maintenance of water resources.

In most developing countries, solid waste disposal is closely linked with urban wastewater drainage in that open sewers and drainage channels often end up also being receptacles for solid wastes generated in the community (Ajayi, 1993). This leads to blockages, reductions in capacity and an exacerbation of flooding problems. In such situations, it is suggested that the institutional arrangements be along the lines of an 'Integrated Environmental Service' provider (incorporating water supply, wastewater and solid waste functions) and a corresponding 'Integrated Environmental Control Agency'.

It is the author's view that the proposed institutional arrangements will provide an appropriate framework for the equitable allocation of resources to the various environmental service needs and should result in a climate conducive to the implementation of the alternative approach being advocated.

The alternative approach

The alternative drainage concept being advocated utiliszes the single pipe system philosophy (Smisson, 1980). This approach differs from the conventional combined drainage concept in that no overflows are permitted from the single pipe system. A single sewer network system conveys the highly polluting urban wastewater sources to a treatment facility prior to its discharge as treated wastewater into a receiving water course. Flows in excess of downstream sewer capacities during rainfall are retained adjacent to the intakes to the sewer system, in local transient storage.

Wastewater that has entered the sewer system is prevented from overflowing or flooding a downstream location because the rate of release of water from upstream parts of the catchment is limited, by the use of flow control devices, to the capacity of the downstream sewer. Details of the basis for design of the single pipe system are described elsewhere (Smisson, 1980).

The single pipe system design philosophy recommends the use of *minor* and *major* drainage systems. The minor system consists of a piped drainage network constructed to serve the area with sufficient capacity to convey base flows and the 'more frequent' storm runoff from roads, highways and other paved areas likely to be sources of relatively polluted runoff.

The major system consists of the natural drainage routes and patterns evolved by nature prior to mankind's interference through development. This is defined by the topography and geomorphology of the area. Overland flow routes for the major system may incorporate roadways, existing streams and their flood plains and suitably graded lawns, park lands and green belts. Overland routes can be engineered such that large parks and gardens etc. are utilized as flow attenuation or retention/detention basins which encourage evapo-transpiration and percolation.

Sadly, the lack of adequate planning policies and controls coupled with a lack of awareness of the importance of the natural drainage routes has resulted in developments that alter or obstruct the natural drainage paths. There is a need therefore for a increased awareness of the impacts of uncontrolled urban developments and a clear demarcation and inclusion of natural drainage paths in urban development plans. Plans for land use changes should incorporate features to increase the surface storage and reduce the velocity of overland flows. The hydrogeological characteristics of the area should be taken into account such that the maximum potential for percolation into the underlying soil is realized.

A review of the hydrological cycle would suggest that one of the key objectives in environmental water quality protection should be that of preventing the contamination of relatively unpolluted water sources. Rain water tends to be the least contaminated of sources. Collection and storage of rainwater water runoff from roofs, for example, could serve the dual purpose of significant reductions in the volume of runoff into sewer networks and the provision of water which could be used for general purposes such as watering of lawns and gardens.

Flooding from combined sewers in most urban centres is caused by increase in runoff rates and volumes resulting from expansion and growth beyond the core area. The search for conventional solutions of larger relief sewers or detention basin in the areas where the problems are manifest (i.e. the urban centre), are fraught with problems of lack of space and congestion of services. Adopting the alternative philosophy of prevention rather than cure would mean solutions investigated look at ways in which flows into and through the urban centres can be reduced or attenuated before they arrive at the problem areas.

The single-pipe system drainage concept advocated as an alternative to the conventional approach, represents a shift from a *curative* approach to a *preventative* approach. This inevitably results in the conservation of resources and leads to cost beneficial schemes for either the provision of new urban drainage infrastructure or the resolution of problems with existing infrastructure caused by the urbanization process. The potential benefits resulting from adopting such an approach are demonstrated by the case studies presented.

Case studies

Three case studies showing the benefits and significant cost savings that accrue from adopting the alternative philosophy of effecting control closer to source in a distributed fashion are described.

York, Toronto — Ontario, Canada

The borough of York , a suburb of Toronto, Canada, has a combined sewer system and in the past had suffered from severe sewer backup and overflows polluting local rivers. In 1968, following a consultant's recommendation, York embarked on a $50 million program along the traditional structural-intensive solution of sewer separation and storm sewer enlargement.. Between 1968 and 1976, York spent an average of $646 000 per annum (22 percent of its annual budget) on this project (GAO, 1979).

By 1976, the borough council had become quite concerned about the tremendous cost of the project and engaged an engineering firm to find an alternative solution. This firm determined that the conventional approach of relief sewers was far too costly and suggested an alternative approach which involved using flow regulators in catch basins, constructing limited-storage underground tanks, and either disconnecting down spouts (from roofs), or installing restrictors in the down spouts. Under this approach, when sewer system capacity is exceeded, stormwater would be temporarily stored in underground tanks or on the surface for slow release into the system.

York opted for a 10-year storm protection and accepted a final cost of $987 633. The alternative approach was completed in 1978, and has worked satisfactorily with no reported flooding.

Wadley Road — Waltham Forest, London

The Wadley Road Storm Sewer System serves a steeply sloping catchment area of approximately 20 Ha. Overflows from storm water sewers had inundated Wadley Road every year for as long as residents can remember creating flooding up to about 1 metre deep (Andoh, 1994).

The only solution which seemed possible (adopting the traditional approach) was the construction of a bypass sewer system at an estimated cost of £90 000 to £100 000. Though this solution would cure the flooding problems at Wadley Road, it run the risk of flooding another street further downstream. A review of the problem showed that a viable, and by far more cost effective alternative, solution would be to use Hydro-Brake™ flow controls to slow down flows and mobilize available system storage throughout the catchment area upstream. Nine Hydro-Brake™ flow controls of various suitable sizes were installed at a cost of £24 000 resulting in a reliable economical solution well below the cost of an unsatisfactory traditional alternative.

City of Evanston, Illinois — USA

Evanston, a community with a population of approximately 75 000 is served by a combined sewer system. Sewer overloading leading to frequent backups, occurring up to six times a year, was a major problem facing the city. In 1987, an engineering consultant was engaged to evaluate the problem and develop a cost effective alleviation program for the City's combined sewer problems (Barber et. al., 1994).

The traditional solution of relief sewers/sewer replacement, was estimated to cost $290 million. In addition, this solution would cause major disruption affecting up to 90 per cent of the City's streets. The high cost of the traditional solution coupled with the potential disruption to local residents caused the city to seek a more affordable solution.

A review of alternatives resulted in the adoption of a plan involving partial sewer separation with above ground storage and overland flow and inlet restrictors installed in catch basins to limit the inflow to the hydraulic capacity of the existing system. This alternative is estimated to cost $143 million, approximately 50 per cent of the conventional sewer relief scheme.

Following completion of the first phase of the project, Evanston has been subjected to several storm events which would have created basement backup in the past. A survey of the area's residents revealed that no backups were experienced.

Concluding remarks

It is more now than ever before in the history of mankind becoming evident that ecological systems cannot cope with many of mankind's activities resulting from industrialization and urbanization.

Unfortunately, mankind has traditionally operated and still to an extent, operates under a feedback law with a control action being sought and implemented only when an undesirable effect or state of affairs is observed. *'Urban drainage practice and control philosophy until recently has, as result, been based on solving localized problems either by transferring excessive flows in drainage systems downstream by upgrading sewer pipes or, relieving localized flooding by constructing local storm overflows'* (Andoh, 1994).

Problems of downstream flooding and pollution and the realization of the interdependence and interaction of the effects of the localized control measures, has focused attention, in more recent times, on the need for an integrated systems approach which looks at urban drainage networks as part of integrated catchment systems incorporating 'flow sources', 'in-sewer' components, 'end of pipe' systems and receiving waters. With an integrated systems approach, the effects of localized control measures on the entire system can be evaluated leading to the evolution of optimal solutions satisfying multi-objective criteria in a holistic manner. When implemented within a framework of sustainable control philosophies, such as the alternative approach being advocated, cost effective solutions are obtained as demonstrated by the case studies presented.

The main factors which have prevented the widespread adoption of the alternative approach are *'Tradition'* and *'Institutional Issues'*. Engineers have been trained to think along the lines of the traditional concepts and the institutional arrangements in most countries are not conducive to the implementation of the alternative approach.

Provision of urban drainage infrastructure along traditional lines is too costly. Developing countries are not going to be able to afford the traditional approach and need to look for cheaper alternative strategies for resolving problems associated with urban wastewater drainage. The alternative approach described provides a framework which should enable staged implementation of effective urban drainage infrastructure.

In order for developing countries to realise the potential of the alternative approach and thereby leap frog the traditional developmental life cycle, there is the need for a *paradigm shift* and a review of institutional arrangements. Changes in institutional structures to reflect integrated environmental service provision and control and an increased awareness through educational programs and public awareness campaigns should help in the evolution of an environment conducive to the implementation of *'Urban Drainage The Natural Way'*.

References

Ajayi, J.O.K. (1993) 'Managing the Lagos metropolitan stormwater runoff problems under structural adjustment economic policy'. In: *Proc. 6th Int. Conf. on Urban Storm Drainage*, J. Marsalek and H.C. Torno (Eds.), Seapoint Pub. Victoria, B.C.

Andoh, R.Y.G. (1994) 'Urban Runoff: Nature Characteristics and Control'. *J. IWEM*, 4, August.

Barber, D., Persaud, R., Stoneback, D., and Velon, J. (1994) 'Evanston's Unique Solutions to Large Combined Sewer Problems.' *IWEA*, March 31.

HRD., (1992) *Urban Drainage: The Natural Way* Hydro Research & Development (pub.). ISBN 0 9521389 0 5.

GAO (1979) 'Large Construction Projects To Correct Combined Sewer Overflows Are Too Costly', report by the Comptroller General of the United States, CED-80-40 December 1979.

Smisson, R.P.M. (1980) 'The Single Pipe System for Stormwater Management'. *Prog. Wat. Tech.*, 13, 203.

Sonuga, F. (1993) 'Challenges of urban drainage in developing countries' In: *Proc. 6th Int. Conf. on Urban Storm Drainage*, J. Marsalek and H.C. Torno (Eds.), Seapoint Pub. Victoria, B.C.

Wright, P. (1992) 'The Impact of the EC Urban Waste Water Treatment Directive' *J. IWEM*, 6, December.

Wastewater disposal and problem soils in Lanzhou, China

Tom Dijkstra, WEDC, UK

LANZHOU CITY is the capital city of Gansu Province, PR China. The city is built on the relatively level Huang He (Yellow River) terraces at an altitude of about 1500m and has an estimated population of about two million. The region is characterized by a hilly to mountainous environment where an undulating bedrock relief is covered by a semi-continuous drape of Quaternary silty aeolian deposits called loess. These loess deposits are subdivided into four stratigraphical units comprising the Wucheng loess (Early to Middle Pleistocene), the Lishi loess (Middle Pleistocene), the Malan loess (Late Pleistocene) and a thin Holocene loess layer (see A.O. Billard *et al.* 1993; Derbyshire *et al.* 1993; Liu and Chang 1964). These loess deposits may reach thicknesses of more than 300m. Malan loess deposits cover extensive parts of the Huang He river terraces and frequently small scale failures in these loess deposits occur within the boundaries of the city.

The region is characterized by an arid to semi-arid climate with annual precipitation rates of 200 to 500mm and a potential evaporation of about 1500mm/yr. As a consequence the ambient field moisture content of the loess deposits are only partially saturated characterized by moisture contents as low as 8 to 10 per cent. These low moisture contents are essential for maintaining the loess structure which consists of a framework of silt sized particles (predominantly quartz and some feldspars) being supported by a cemented network of clay-sized particles forming bridges and coatings. The presence of cementation determines to a large extent the structural strength of the deposits. The brittle cementation bonding is a function of the availability of calcium carbonates and, to a limited degree, also soluble salts may contribute to this brittle cement. Clay minerals can play an important role in forming ductile bonds dispersed through the openwork fabric of loess. However, the clay mineral content of the non-weathered aeolian loess is low and

under low partial saturation loess fails in a brittle mode. On average, Malan loess contains about 11 per cent calcium carbonates, while in the Wucheng loess it may be as high as 16 per cent. For the Malan loess its open work structure has characteristic void ratios greater then 1.0 (indicating that the volume of voids is larger than the volume of solids). The occurrence of the widespread carbonate cementation gives the loess its high structural strength which can be summarized by using the effective cohesion and internal friction angles for the undisturbed samples of 30 kPa and 26⁰ for the Malan loess, 54 kPa and 30⁰ for the Lishi loess, and 52 kPa and 35⁰ for the Wucheng loess (Table 1; see Derbyshire *et al.* 1993, 1994; Dijkstra *et al.* 1994, and Wang *et al.* 1991).

The highest mountains around Lanzhou city reach heights of 2600m to 2900m generally having long steep slopes characterized by a relative relief ranging from 200 to more than 500m and slope angles as high as 45⁰. Expansion of the city is thus severely limited. A current need for more space for industrial and urban expansion of the city, intensified by the greater attention the central government in Beijing is paying to formerly 'marginal zones', has led to extensive levelling activities just north of the city in the valley of the Luo Guo Gou in order to create space for the establishment of new industry and some housing. However, much still depends on an increase in housing density in certain areas of the city and the utilization of steep, potentially unstable escarpments of the river terrace edges and the lower slopes of the loess-covered surrounding hills where the houses are built on man-made terraces dug into the loess deposits. The nature of these loess deposits allows for the creation of caves which are used for storage or even housing of entire families. Slope stability is, of course, severely affected by the digging of these caves. However, another, arguably more important, factor is the way in which the local households have to dispose of their wastewater. Al-

Table 1. Some representative properties of Malan loess samples

Loess type	s.g. [10³kg/m³]	r. [10³kg/m³]	m.c. [%]	e_0	n	LL	PL	PI	c [kPa]	ϕ
Wucheng	2.75	1.74	7.51	0.699	0.41	27.0	16.6	10.4	52	33⁰
Lishi	2.78	1.60	9.65	0.905	0.47	31.1	17.8	13.3	54	30⁰
Malan	2.72	1.38	6.12	1.092	0.52	28.4	17.0	11.4	30	26⁰

s.g. = specific gravity; r. = bulk density; m.c. = moisture content; e_0 = void ratio; n = porosity; LL = liquid limit; PL = plastic limit; PI = plasticity index, c = cohesion, ö = internal friction angle.

though the city is situated in an arid to semi-arid environment sufficient water can be used from the Huang He and from water bearing strata to the north of the city. At present there is a continuous water supply to the majority of the suburbs of Lanzhou. However, facilities for the disposal of waste water are far from adequate in most parts of these suburbs and waste water disposal frequently occurs by letting it drain away into the loess, directly adjacent to the houses. Although part of this water will rapidly evaporate during the warm summer months, sufficient amounts of water slowly seep through the loess deposits causing a gradual destruction of the loess fabric and a decrease of the structural strength. Where a waste water drainage infrastructure is available poor maintenance combined with frequent small-scale slope movements has led to numerous leaks in this systems. At these sites rapid disintegration of the loess fabric occurs, followed by piping or even mass failure of the slopes.

In the next section, the effects of water on loess strength will be discussed followed by an assessment of changing slope stability conditions of a slope in Malan loess within the city limits of Lanzhou.

As an example of the impact of human activity on slope instability conditions within the city of Lanzhou a landslide near Baita Park will be discussed. This landslide took place in one of the densely populated suburbs of Lanzhou city on the 9 November 1986 and, despite its limited size, caused the death of seven people and the destruction of several houses. A dispute has arisen about who should be held responsible for the initiation of the slope failure. On the one hand, movement may have been initiated by extensive weathering of the loess caused by seepage of waste water sewage from the houses on the higher slopes of the terrace. Alternatively, slope stability may have been reduced sufficiently by uncontrolled excavation activities at the base of the slopes (including local oversteepening of the slope and the construction of a loess cave for storage purposes). The stability analysis presented here does not settle this question and merely illustrates the results of the geotechnical testing in relation to the increased moisture content and the potential effects of a long term throughflow at this site. The Baita Shan profile has been selected for three reasons. First, reliable data are available for the reconstruction of both the pre- and the post-failure profiles. Second, the slope is representative of large areas of terrace edge in the built-up area of Lanzhou, and the instability of such slopes is known to cause extensive damage to properties and to threaten lives and livelihoods in this densely populated area. Third, because the slope is homogeneous it is unlikely that internal anisotropy (such as palaeosols or changes in the type of loess) predetermines the location of failure surfaces, or that a build-up of a ground water table will occur within the slide mass. A number of field observations indicate that progressive failure of these slopes is a common phenomenon. These indicators include: extensive cracking of the upper slopes; salt efflorescences at spring zones; localized subsidence on the terrace levels associated with the collapse of the loess fabric which, in the case of Malan loess, may amount to

a linear decrease as much as 15 per cent; and steps in the terrain associated with large scale creep along potential failure planes (cf. CSCCC 1978 and Dijkstra et al. 1994).

The variations in loess structural strength derived from extensive geotechnical testing in the laboratory and in situ can be used to assess the effects of progressive weathering and varying moisture contents on the failure of loess slopes in Gansu Province. The changes in slope stability at this site are presented below using slope stability analyses for a range of 'weathering stages' which are thought to represent the effects of progressive weathering of the loess deposits at this site comprising the dissolving of the readily soluble components of the loess fabric and the effects of changing pore pressures.

When tested under low normal stresses, peak strength conditions of the Malan loess are frequently observed. However, at higher normal stresses this peak strength condition tends not to develop and failure of the sample is characterized by strain hardening resulting of the plastic deformation in the shear zone, whereby the fabric is changing as a result of a rearrangement of the particles. Subsequently, the more effective particle interlocking induces a slight increase in the effective angle of internal friction.

A large difference exists between undisturbed Malan loess (characterised by an extreme openwork fabric), and the same sample after shearing. The development of a more compact fabric, stress anisotropy, and discontinuous shear zones severely alter the behaviour of this type of loess. Permeability values, collapsibility coefficients and void ratios decrease sharply.

A set of samples, derived from those used for the undisturbed loess tests, was used to remould manually and tested in a modified Bromhead ring shear apparatus (see Bromhead 1979 and Boyce et al. 1988). The tests were carried out to assess the effects of changing moisture contents on effective (apparent) cohesion and effective angle of internal friction. It was found that the apparent cohesion gradually increases to a value which is about 40 to 50 per cent of the effective cohesion values of the undisturbed samples tested at their unaltered field mois-

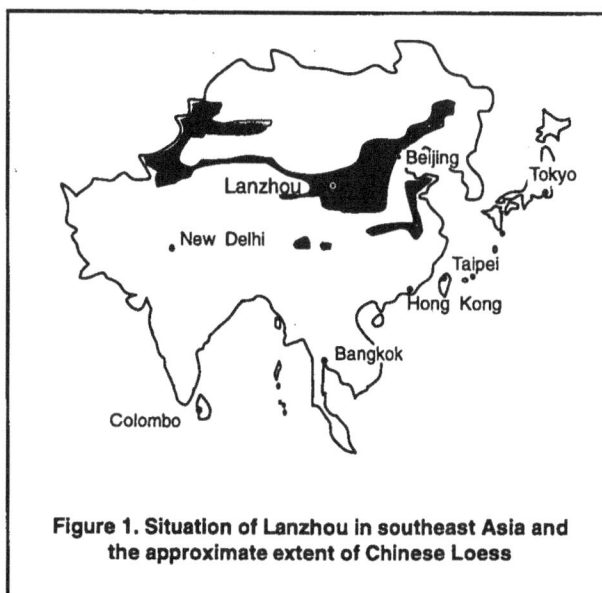

Figure 1. Situation of Lanzhou in southeast Asia and the approximate extent of Chinese Loess

Table 2. Malan loess strength data used for the slope stability analyses of Baita Shan

	w [%]	tan ϕ'_r	c'_r	WP
x_1	5	0.47	45	0
x_2	5	0.78	6.2	0.25
x_3	10	0.75	10.6	0.50
x_4	15	0.72	16.2	0.75
x_5	20	0.7	5	1.00

Weathering percentage (WP) is the degree to which a slice has been affected by weathering.

ture content of around seven per cent. When the moisture content of the remoulded samples exceeded a threshold value of about 18 to 20 per cent, a sudden decrease was noted in the effective cohesion as well as an increase in the scatter of the data occurs. The effective angle of internal friction indicates that the gradual increase in the thickness of the water membranes surrounding the particles plays an important role in the increase of the internal friction angle in the moisture content trajectory between two and five per cent. At moisture contents greater than five per cent a gradual decrease of the effective internal friction angle is noted, with again a greater scatter of the data occurring at moisture content values exceeding 18 to 20 per cent which are just exceeding the plastic limit and are representative of saturation degrees in excess of 0.95.

Slope stability analysis was carried out using a limiting equilibrium method of slices following the method of Sarma (see e.g. Sarma 1973, 1979). In our example the slope mass was divided into 14 slices. Three separate calculation series were used to simulate stepwise internal weathering along the known critical slip surface from totally undisturbed to fully remoulded and weathered (see Figure 2). In these series of calculations weathering was assumed to take place either from the top or from the toe of the slope. In the third series weathering was assumed to progress alternately from the toe of the slope and from the top of the slope. Based on the data obtained from the geotechnical tests of both the undisturbed and manually remoulded samples, an array of strength characteristics was constructed and used as a surrogate for the various stages simulating progressive weathering in loess (Table 2). According to these data the hypothetical progressive weathering is simulated by changing the c' and tan phi' values from peak values into remoulded values for one or more of the fourteen slices. It is assumed that each slice behind the 'weathering front' is in a remoulded state and undergoes a progressive increase in moisture content (5, 10, 15 and 20 per cent). Additionally, it is assumed that the sliding body is dry (moisture content not exceeding five per cent), and that near-saturation conditions occur only at the slip surface. The results

of the analysis are expressed as the equivalent percentages (eq) of the total slip surface length influenced by weathering. For this purpose the weathering percentages are used as mentioned above and described in Table 2. Typically, the weathering situation at any time was defined with the following relationship:

$$eq = a + 0.75b + 0.5c + 0.25d \qquad (1)$$

where
a = percentage of the (remoulded) slip surface at 5% moisture content (x_2)
b = percentage of the (remoulded) slip surface at 10% moisture content (x_3)
c = percentage of the (remoulded) slip surface at 15% moisture content (x_4)
d = percentage of the (remoulded) slip surface at 20% moisture content (x_5)

The results of the analyses are shown in Figure 2. With the weathering from the top of the slope a strong initial decrease of slope stability occurs, which is explained by the considerable drop in effective cohesion values when the loess changes from an undisturbed into a weathered ('remoulded') state and the relatively small effective normal stress acting on the failure plane at this point. This effect gradually becomes less important with the increase in the effective internal friction angle taking effect at greater depths after further progression of the internal weathering of the slope. Slope stability analyses carried out while adding weathered slices from the toe of the slope indicate an initial increase in factor of safety due to the higher internal friction angles of the weathered slices causing the lowest slices to act as buttresses against movement of the upper parts of the slopes. After adding the third slice this effect is diminished by the changes in the effective apparent cohesion values, leading to a sudden drop in the safety factor. This curve does not follow the relatively smooth progression of weathering from the top, but for certain additional slices the factor of safety increases above the previous value. The reasons for this are thought to be a combination of the complexity of strength parameter variation (c' falls, rises and then falls again while tan phi' rises then falls), the irregular pre-failure surface profile and the considerable influence of effective normal stress up to at least 50 per cent weathering, and the rapid inclination of the failure plane.

During the simulation of the combined effects of weathering of the top and the toe of the slope an initial increase in factor of safety is observed, which is immediately followed by a rapid decrease and a subsequent small increase. Both increases are caused by the higher values of the effective internal friction angle of the remoulded loess. These higher values are particularly effective in the lower part of the slide, where the effective internal friction angles are significantly higher than the local slip surface angle measured at the base of each slice and the effective normal stress is greatest. The effect will be even more pronounced for larger mass movements where the changes in effective internal friction become relatively more im-

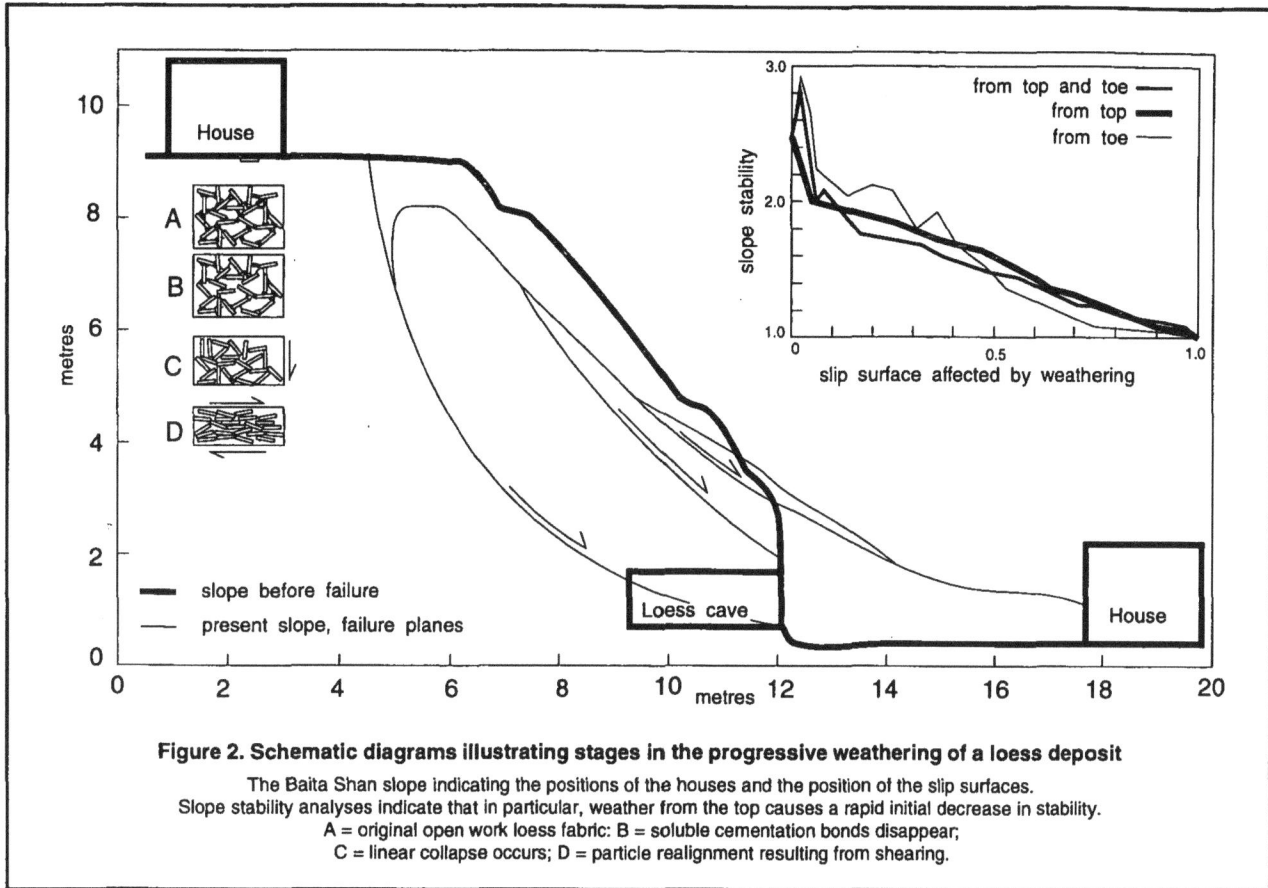

Figure 2. Schematic diagrams illustrating stages in the progressive weathering of a loess deposit
The Baita Shan slope indicating the positions of the houses and the position of the slip surfaces.
Slope stability analyses indicate that in particular, weather from the top causes a rapid initial decrease in stability.
A = original open work loess fabric: B = soluble cementation bonds disappear;
C = linear collapse occurs; D = particle realignment resulting from shearing.

portant than the changes in effective cohesion. Figure 2 indicates that slopes which, in an unweathered state, have safety factors close to unity might fail after only a short period of progressive weathering. This process will take considerably longer for the initially more stable slopes. Within the urban environment the man-made cut slopes are frequently close to equilibrium and therefore moderate amounts of weathering affected by e.g. the uncontrolled disposal of waste water may bear great risks for these sites.

The following conclusions may be drawn from the results and discussion presented here.

1. The youngest (Malan) loess has a peak strength at low normal effective stresses but fails plastically at normal effective stresses above approximately 100 kPa. The older (Lishi and Wucheng) loesses exhibit greater strength and brittle failure because they are cemented.

2. The behaviour of the remoulded loess is markedly different showing higher effective internal friction angles and lower effective cohesion values. The change in the effective angle of internal friction is caused by the redistribution of the particles into a denser packing: the loss of cohesion is caused by the breaking up of the cementation bonds.

3. Weathering of loess, for example by dissolution of the cementing bonds, results initially in structural collapse and a consequent large increase in effective angle of friction and reduction in effective apparent cohesion.

4. As the water content of the remoulded loess rises, the effective apparent cohesion rises proportionately for all three loess types up to a limiting value of approximately

20 kPa at a water content of 16 to 18 per cent, after which the value falls rapidly to a much lower, somewhat variable level.

5. The effective angle of internal friction rises to a maximum at a water content of approximately five per cent and thereafter falls to a lower value at approximately 20 per cent, whereupon the effective strength parameters stabilize since the soil becomes completely saturated.

6. Progressive weathering from the top of the loess slopes causes a dramatic reduction in their stability, whereas progressive weathering from the base results initially in stabilization of these relatively large slopes because on collapse of the loess structure of these Malan loess deposits a rise in the effective internal friction angle occurs. Detrimental effects only occur once the weathering has progressed beyond a certain stage.

7. The processes described above are not only related to loess soils, but are representative of a wide range of cemented and/or collapsible soils which cover large parts of the globe. Occurring predominantly in arid to semi-arid zones these soils maintain their open structure under low moisture contents and are therefore extremely sensitive to changes in their internal hydrological regimes. Particularly in an urban environment relatively small-scale hazards bear a great risk and have a profound impact on human activities.

This paper presents some of the results taken from a much larger study of landslides and mass flowage in the loess of north-central China supported by the Council of the European Communities and the Government of Gansu Province, P.R. China and carried out in 1987-1993. Special acknowledgements are due to my Chinese research colleagues of the Geological Hazards Research Institute in Lanzhou, and to the other universities and research institutions participating in the research.

References

Billard, A., Muxart, T., Derbyshire, E., Wang, J.T., and Dijkstra, T.A. 'Landsliding and land use in the loess of Gansu Province, China.' *Zeitschrift für Geomorphologie, Supplement Band* **87**, 117-131, 1993.

Boyce, J.R., Anayi, J.T. and Rogers, C.D.F. 'Residual strength of soils at low normal stresses.' In: Bonnard, C. (Ed.) Landslides. *Proceedings Fifth International Symposium on Landslides*, Lausanne, Switzerland, **1**, 85-88, 1988.

Bromhead, E.N. 'A simple ring shear apparatus.' *Ground Engineering*, **15(5)**, 40-44, 1979.

CSCCC (Chinese State Commission on Capital Construction) *Code on the building construction in regions of collapsible loess (TJ 25-78)*. China Building Industry Press, Beijing (in Chinese), 1978.

Derbyshire, E., Dijkstra, T.A., Billard, A., Muxart, T. Smalley, I.J. and Li, Y.J. 'Thresholds in a sensitive landscape: the loess region of Central China.' In: Thomas, D.S.G. and Allison, R.J. (Eds.) *Landscape sensitivity*. John Wiley and Sons Ltd, Chichester, 97-127, 1993.

Derbyshire, E. Dijkstra, T.A., Smalley, I.J., Rogers, C.D.F, and Li, Y.J.. 'Failure mechanisms in loess and the effects of moisture content changes on remoulded strength.' *Quaternary international*, 1994, in press.

Dijkstra, T.A., Rogers, C.D.F, Smalley, I.J., Derbyshire, E., Li, Y.J., and Meng, X.M.. 'The loess of North-central China: geotechnical properties and their relation to slope stability.' *Engineering Geology*, 1994 in press.

Liu, T.S., and Chang, T.H. 'The Huangtu (loess) of China', *Report of the 6th INQUA congress*, **4**, 503-524, 1964.

Sarma, S.K. 'Stability analysis of embankments and slopes.' *Géotechnique*, 23, 3, 423-433, 1973.

Sarma, S.K. 'Stability analysis of embankments and slopes.' *Journal of the Geotechnical Engineering Division*, ASCE, GT 12, 15068, 1511-1524, 1979.

Wang, J., Derbyshire, E., Meng, X., Ma, J. 'Natural hazards and geological processes: an introduction to the history of natural hazards in Gansu Province, China.' In: Liu Tungsheng (ed.), *Quaternary Geology and Environment in China*, Papers submitted to XIII INQUA 1991, 285-296, 1991.

Sanitary aspects of canal project

M. S. Fernando, Sri Lanka Land Reclamation & Development Corporation, Sri Lanka

THE CANAL NETWORK in and around the capital of Sri Lanka, Colombo, is in such a deteriorated condition that it can only be compared to a time bomb about to be exploded. Canal water is so polluted that in most places the canals act as anaerobic ponds. This unhealthy condition is further aggravated by the low income communities living along canal banks. The only ray of hope in this dark world is the *Canal Rehabilitation Project*, the implementation of which assures a healthier environment for those living in close proximity to canals.

This paper deals with the present situation of the canal system with relevance to sanitation and the implementation of the canal project to improve the situation.

Canal Project

The Greater Colombo Flood Control and Environment Improvement Project, termed here as the *Canal Project*, is concerned with rehabilitating the canal network and constructing underground drainage systems, within Colombo and suburbs, for reducing the frequency of flooding. phase I of the project, currently being implemented, deals with the rehabilitation of the canal network and Phase II of the project, which will be implemented shortly, will deal with the construction of surface drainage schemes, most of which will be underground. Phase I of the project' considered a major project in Sri Lanka, is being implemented by the Sri Lanka Land Reclamation and Development Corporation at a cost of 3.61 billion rupees. (About 74 million US dollars). The project was launched in September 1993 and the work is expected to be completed in July 1997.

The project area covers 85.7 sq km comprised of the Colombo Municipal area and parts of other local authorities, which fall within the Greater Colombo area. The project area is bounded on the west by the sea, on the north by the left bank flood bund of the Kelani Ganga, on the east by a line running from Ambatale in the north to Kottawa in the south, and the southern boundary is along a line running from Mt. Lavinia to Kottawa.

Phase II of the Canal Project will address the problem of isolated flooding of areas which are not connected to the canal network due to the absence of proper drainage systems. In addition, pollution aspects of the canal network will be studied under this project. The anticipated cost of the project is Rs. 2.33 billions (47 million US dollars). It is expected to commence this phase of the Canal Project in February 1995 and complete in September 1999. The project area includes Dehiwala - Mt. Lavinia Municipal Council area and Urban council areas of Kotte, Moratuwa and Kolonnawa, in addition to the phase I area.

Sanitation aspects of the project

The sanitation aspects of the canal project are discussed here under following topics:

1. Pollution of water in the canal network and associated health problems.
2. Sanitation aspects associated with the low income communities living along the canal banks.

Canal water pollution

The pollution of canal water can be attributed to factors such as:

- Industrial effluent discharged to canals.
- Wastewater and sewage directly discharged to canals.
- Dumping of garbage.
- Some low income communities defecate on canal banks, open lands etc which end up in the canal system during the rainy season.
- Chemicals and other wastes from agricultural lands.

As a result of the above factors, the canals have become so polluted that in many places they act as anaerobic ponds, which are too shallow to allow any aerobic or facultative development.

According to the Beira Lake Restoration Study (1993), the data collected by the Central Environmental Authority during a period of 23 months (from March 1991 to February 1993) have shown that St. Sebastian canal had the following characteristics with regard to the water quality:

'High turbidity, higher BOD and COD, richer in nutrients, reduction of nitrate into nitrite and ammonia (due to eutrophication), high concentrations of metals and faecal coliform.'

The above highlights the extent of pollution in canals.

The effects due to pollution can be categorized as:
- effect on human health by bathing and washing in the polluted water.
- effect on human health from mosquito-borne diseases associated with stagnant or low flow water.
- accumulation of toxic pollutants in fish and other aquatic organisms.
- accumulation of toxins in the food chain.
- ground water degradation.

It is important to consider the impact due to water pollution from current and future industries and other sources to formulate strategies to control in an efficient and economical manner consistent with the development needs of the country. This remains a formidable challenge for the Canal Project.

Polluting industries

The effluent discharge by industries remains as a major contributory factor in water pollution. With the Inspector of Factories of the Labour Department, 4 800 industries have been registered in Colombo Urban Area (Colombo Urban Area is the most urbanized area within Greater Colombo). Through its Environment Protection Licensing Scheme, the Central Environment Authority has identified 365 of these industries as having a pollution potential. ('Environmental Management Strategy for Colombo Urban Area', Volume 1, March 1994). In addition, there is another concentration of polluting industries in Dehiwala - Mt. Lavinia area discharging pollutants to canals.

Most of these industries are located in unplanned areas, which have evolved informally as areas of concentrated industrial activity. This situation, combined with the fact that few industries have adequate pollution control systems, constitutes a potential threat to the environment and public health. The effluent from most of these industries ultimately ends up in water courses causing water pollution.

Dumping of garbage by people living close to canals, specially by the low income communities, is also another problem associated with sanitation. This aspect will be dealt with under the low income communities.

Discharge of sewage

In Greater Colombo area 1.7 million people are using on-site sewage disposal systems, mainly septic tanks. (Greater Colombo wastewater and sanitation master plan, executive summary, 1993). However, some of these do not operate properly due to:

○ poor design or construction
○ incompatible soil conditions
○ high water table
○ overloading
○ inadequate maintenance
○ reduction of porosity due to deposition of fine particles

When problems such as those mentioned above arise, many connect effluent to water courses, which ends up in the canal system.

There are over 350 000 people living in slums, shanties and other semi-permanent dwellings within the Greater Colombo area (Greater Colombo wastewater and sanitation master plan, 1993). Most of these people have no sanitation facilities and defecate on canal banks and other open spaces, most of which end up in the canal network during the rainy season. It is a common sight to see raw

sewage floating in canals, sometimes in polythene bags. Due to this, the canal labourers are reluctant to get into water for maintenance purposes and skin and other diseases are common among canal labourers. Even those who have toilets do not have proper septic tanks and soakage pits and the outlets are usually connected to canals. It is not only the shanty population, living by the canal, that contributes to this pollution, government and private institutions also connect wastewater and sewer outlets to canals.

The water pollution due to agriculture is also significant in the canal network. The Environmental Management Strategy report states that there are over 20 000 ha. of land in Colombo Urban Area still used for agriculture. ('Environmental Management Strategy for Colombo Urban Area', Volume 1, 1994). The organic wastes and the agrochemicals from these agricultural lands contribute significantly to a deterioration in the water quality.

Mosquito problem

Mosquito breeding is a serious problem in the project area which can be attributed mainly to the low flow or stagnation of water in canals. According to the report prepared by the Inter-Agency Committee appointed by the Government of Sri Lanka for the control of Vector and Nuisance Mosquitos, there were 140 species of mosquitos identified and recorded in Sri Lanka in 1987. The diseases transmitted by mosquitos are malaria, urban filariasis, dengue, dengue haemorrhagic, yellow fever and japanese encephalitis. Those other than the vector mosquitos, although not transmitting diseases, are a serious nuisance to the residents in the project area.

After the completion of the project, the drainage capabilities of the canals will be improved. In addition, arrangements have been made to have a flushing system of the canal network during the dry weather. All these measures will contribute to overcome the problem of stagnation of water. Therefore, after the completion of the project, a vast relief in mosquito menace could be expected.

Other health problems

Water sanitation related diseases are a major health problem associated with the canal network. Notifiable diseases within the water sanitation-related category include: typhoid/para-typhoid fever, dysentery, viral hepatitis, dengue haemorrhagic fever, japanese encephalitis, leptospirosis, malaria, filaria, cholera, and poliomyelitis.

Recorded data from 1965 to 1990 confirm that diarrhoeal diseases are the major cause of morbidity in Sri Lanka. Although intestinal infections have shown a declining trend since 1965, this group of diseases ranked as the number 1 morbidity factor with the associated highest mortality. (Greater Colombo wastewater and sanitation master plan, 1993). These diseases still remain as major health problems in Sri Lanka. This situation could vastly improve in Colombo and suburbs if adequate pollution control measures are taken with regard to the canal system.

A detailed study on canal pollution is envisaged, under the phase II of the Canal Project, in order to identify the action programme to improve the situation. The relocation and upgrading of low income communities, which is being carried out under the project, is a giant step towards pollution control. In addition, the Central Environment Authority is taking action to   control the discharge of effluent to water bodies from industries.

Low-income communities in canal banks

The rising value of land over the last 50 years has pushed the low income communities to least favourable lands. The only way such communities were able to maintain the location necessary for their survival was to build their homes on the canal banks or on similar marginal lands. These low income communities are characterized by low levels of income, high density of housing, lack of physical infrastructure amenities and insecure land tenure.

Some of these low income communities have constructed dwellings along the canal banks, sometimes within the water way itself obstructing the water flow, making the maintenance work of canals difficult due to the restrictions on access to the canals. This also poses problems for the rehabilitation work as the space required for widening canals and the use of construction equipment is encroached upon by the low income families.

These low income communities have contributed in many ways the pollution of canal water, which has not only affected their own health but that of others who are living in the vicinity of canals. Therefore, the relocation and upgrading of these low income communities is of vital necessity for the rehabilitation of canals. This part of the canal project has already started and will be completed shortly.

There are many sanitary aspects concerning the low income communities. Although standpipes are available, they are usually located away from these communities and canal water is used for bathing, laundry etc. The canal water is so polluted that this practice contributes to poor sanitary conditions. Due to the unsatisfactory state of canals, flooding is frequent and the houses are inundated with polluted water during flooding, which is contributing to the spread of water borne diseases. The victims of this are mostly children.

Solid waste is dumped into canals due to a number of reasons. There are no ways to dispose of solid waste in most areas. Even the places where solid waste disposal systems exist, they require transit arrangements: mobile steel drums and road side cement structures. The frequency of transit is so low that solid waste deteriorate and emits offensive smells. Dwellers close to canals find it difficult to carry solid waste up to transit points due to heavy congestion of housing units. On the other hand, dumping in the canal is easier and what is dumped is

visibly taken away by the water flow clearing the guilty conscience of those throwing the garbage.

The Sri Lanka Land Reclamation and Development Corporation adopted a two way strategy to solve the problems due to encroachments by shanty dwellers. These consisted of:

- Relocating families in selected lands which are provided with infrastructure facilities.
- Onsite upgrading of remaining families in such a way not to have adverse effects on the canal rehabilitation.

These are discussed below.

It was estimated that there were more than 11 000 families living along canal banks and retention areas at the commencement of the canal project. Out of these 11 000 almost 7 500 would be directly affected by the project.

Therefore, the success of the Canal Project largely depends on the relocation and upgrading of low income communities from the canal banks. Although almost all these dwellers are illegally occupying the canal banks it is not desirable to remove them without their co-operation due to humanitarian reasons. Therefore, this is a complex and extremely difficult task involving many considerations and strong political leadership.

Relocation of low income communities

Sri Lanka Land Reclamation and Development Corporation is implementing the relocation programme with the assistance of the National Housing Development Authority (NHDA). The most important principle followed in this work is to relocate low income families close to their present locations in order not to disturb their day-today activities. The Corporation has developed few large relocation sites in addition to several small sites for this purpose. The persons to be relocated have to be given an attractive financial package to promote voluntary relocation.

Every attempt is made to provide basic amenities such as potable water, electricity, toilets, access roads and other facilities required for a healthy living in these relocation sites.

Generally, whenever sufficient land is available on the canal bank, after meeting all the requirements of the project, on-site upgrading is implemented. The upgrading process involves improvement to low income housing units and provision of basic amenities such as potable water supply, access roads, effective garbage disposal and appropriate sewage disposal. The families are provided with loans to improve their housing units.

Provision of sanitary infrastructure at relocation sites

As the relocating agency, the Sri Lanka Land Reclamation and Development Corporation, has a major role to play with regard to the provision of infrastructure as the communities involved are not capable of providing those themselves. The sanitation facilities, which are made available at relocation sites by the Corporation, are discussed below.

Water supply

Every attempt is made to provide pipe-borne water at relocation sites in the form of standposts, a connection for about 10 families. Most parts of these standposts are buried inside the structure leaving only the tap and the outlet visible. This arrangement makes it difficult to remove parts, which is a common occurrence in low income areas. However, it could be seen that even this arrangement has not discouraged people from damaging standposts in some places. Standposts are always provided with masonry aprons directing water to drains.

One of the problems presently being experienced in some locations is the low pressure of the system. This has resulted in limiting water supply to a few hours a day in some places. The low pressure is also one of the reasons for damaging standposts as the water could be easily collected from the bottom of the standpost once damaged. Another approach taken in areas where low pressures exist is to construct in such a way to lower the elevation of standposts.

At present, the cost of water is borne by the Corporation. In future, action is necessary to hand over the responsibility of selling water to Community Development Councils as neither the Corporation nor any other public sector organization can foot the bill for long due to limitations of financial resources. However, this could be done only after taking appropriate action to improve the income level of the people as most are well below the poverty line and therefore, can not afford to pay for services. At present, the National Housing Development Authority and the National Water Supply and Drainage Board are developing a scheme for low cost connections, where one meter supplies to a block of houses.

Storm water and waste water disposal

The allocation of plots for housing is done in such a way that there would be a sufficient gap between rows of houses at the rear. This gap is used for the construction of drains to carry waste water and storm water. It is necessary to educate the communities to maintain these drains properly as they are earthen drains. Otherwise, these could add more sanitary problems due to stagnation of water as a result of blockage. Many drains are not properly maintained presently as the communities expects this to be done by the Corporation.

Generally all the relocated areas have drainage canals surrounding them, which are constructed for:

- Draining of storm water collected within relocated sites and other adjacent areas.
- Isolating relocation sites from the surrounding areas.

These drainage canals are connected to the canal network in order to have an effective drainage system functioning. The maintenance of these are normally outside the ability of communities due to high costs.

Open drains are found to be most effective in low income community areas due to low cost of construction and maintenance and the ease of maintenance.

Domestic sewage disposal

At relocation sites temporary toilets are constructed in such a way that one toilet can be shared by about six housing units. (The word temporary is used here to indicate that these would be connected to the sewage network in the future.) Usually, four to 10 toilets are constructed in a row and each two are connected to a septic tank. The septic tank used is a circular precast tank with a capacity of 2800 litres. The effluent from the septic tank is led to a soakage pit, generally of the size 6000 litres.

The above sewage disposal system is sometimes not functioning properly as the soakage into the ground is not sufficient specially during the wet weather. Therefore, it is often necessary to get the gully emptiers of the Colombo Municipal Council to remove the sludge. In some locations, dispersion trenches have been constructed.

This unsatisfactory state of affairs with regard to the soakage could be prevented once these toilets are connected to a sewer network.

The Corporation has recently started constructing biological filters through which the effluent is passed to nearby canals. These are circular pipes packed with layers of graded metal. It is yet to assess the performance of these biological filters.

Solid waste disposal

Masonry bins are provided of the capacity of about five cubic metres for the temporary storage of solid waste. They are constructed of bricks and cement rendered. Waste is stored on cement rendered floor. There are no covers for these bins. Each bin serves about 25 families. The collection of the solid waste is usually done by the local authority.

The front end of this bin is normally open to facilitate the collection by lorries. However, it was noticed that the dogs can enter through this opening and scatter the garbage. The community development councils of some sites (e.g. Obeysekera Pura) came out with the solution of constructing wooden barriers which can be removed at the time of collecting the garbage. This is an example where the interest and the active participation of the communities can be used for their own benefit. It is to be noticed here that most of the families living in relocation sites used to dump solid waste into canals earlier. This highlights the fact that when given opportunities, these families also could act as any other responsible citizens.

The approach

The provision of sanitation infrastructure for low income communities associated with the Canal Project has to be done in such a way that the role of the government should be minimal, it should be more affordable both to users and providers and more sustainable in the long run. This is because the provider, Sri Lanka Land Reclamation and Development Corporation is not an organization set up to cater for community works and the major objective in the project is to rehabilitate the canal network. Therefore, with regard to the project the significance of the community works of low income communities is of secondary nature.

The approach in the project up to now has been for the provider to supply all infrastructure, with little or no contribution from the user. But the long term approach should be one of minimal intervention by the public sector and to encourage and assist the community to identify their needs and take necessary action to satisfy them. However, studies carried out indicate that the income levels of the low income communities are very low. (Report prepared by Indeman ltd. on Community Development for Priority Projects, 1993). Therefore, it may not be appropriate to expect a substantial contribution from the community without improving their income levels. Actions that could be taken in this regard are conducting awareness programmes, skill development and introducing small scale income generating ventures. Already some non governmental organizations are active in this area and time is needed to reach harvest in this endeavour.

Another action that can be taken is to have staged development to affordable standards at each stage. This reduces the initial high cost of providing for the infrastructure. For instance, only the major canals can be provided initially, and internal drains could be provided later. However, this approach has to be made carefully, otherwise there could be resentment among the users.

Another important aspect to be noted is that the mere provision of sanitary infrastructure does not improve the living standards in terms of sanitation. It is necessary to educate and cultivate interest in communities in using these facilities. For instance, although the sufficient number of toilets are provided most of the children do not use them and parents, specially mothers, do not persuade them to use toilets. Therefore, worm infections are not reduced much even in communities provided with sufficient sanitary facilities. Another example is that although the solid waste disposal system is vastly improved at relocation sites, still there are some persons preferring to throw garbage to canals because they are unwilling to carry them to transit points.

People who have been living in slums and shanties for a long time without basic infrastructure facilities are sometimes indifferent to family health, preventive health care and use of sanitary facilities. Therefore, it is necessary to organize awareness programmes with the assistance of community development councils. As the major user group of water and sanitation facilities of these settlements are women and children, most of these programmes should be aimed at them. It is very interesting to note that

in Sri Lanka women take active interest in community work. Some community development councils are even headed by women. The high literacy rate is also a plus factor in Sri Lanka in community development work.

Things went wrong

The implementation of a project of this magnitude is not easy. Finding solutions to engineering problems are relatively easy when compared with the complex nature of issues involved with community development work. The low income communities with diversified ideas fuelled by different political opinions make it difficult to accomplish some objectives of the exercise. The Corporation has not been able to convince all the low income communities associated with the project about the benefits that they and the country can achieve through the implementation of the project. Some sections of the communities still believe that the Corporation should supply them with more facilities ignoring the fact that they unlawfully occupied the canal banks earlier and that the life they spent was unhygienic without basic sanitary facilities. Perhaps, due to human nature, it may not be possible to convince all concerned. However, due to this reason the co-operation of the low income communities is not extended fully to make the project more successful.

Conclusions

Implementation of the Canal Project is fulfilling a long felt need in sanitation with respect to the canal network in and around Colombo. This timely step could arrest a potential health hazard exploding to cause significant damage to residents in the vicinity of canals.

Provision of infrastructure related to sanitation is essential at relocation and upgrading sites under the canal project in order to promote healthy living conditions. However, it may not be possible in the future for the public sector to bear this cost. The best possible solution lies in obtaining the active participation of the communities as well as non governmental organizations. However, it is of vital importance that action should be taken to improve the income level of persons living in low income communities in order to pave a way for them to finance their own facilities. Unlike in many other developing countries, Sri Lanka has the advantage that most of the members of these low income communities are educated and therefore, it will be relatively easy to obtain their contribution in these matters.

There is a school of thought that projects of this nature do not give sufficient emphasis to community and sanitation aspects because the decision makers are mostly engineers (S. Niedrum, 1993). This criticism can be overcome by including experts in other disciplines in the decision making team. This also highlights the necessity that the engineering education in developing countries should be more opened introducing other vital aspects such as community development, sanitation and environment.

Acknowledgement

Author wishes to thank the Chairman, Sri Lanka Land Reclamation and Development Corporation for the encouragement given and for granting approval to present this paper. Other officials of the Corporation and the staff of the W.S. Atkins International who supplied with information and contributed in many ways are kindly remembered for their assistance.

References

Abeysirigunawardane H., 'Review of Engineering standards and practices in Sri Lanka relevant to low-cost sanitation', Proceedings of the National Seminar / Workshop on sanitation infrastructure in low income housing projects, National Housing Development Authority, December 1989, pp 11-15.

Beira Lake Restoration Study, Report prepared by Roche International, Urban Development authority, December 1993, pp 64-70.

'Community Development for Priority Projects', Greater Colombo Flood Control and Environment Improvement Project (Phase II), A report prepared by Indeman Consultants (Pvt.) Limited, December 1993, pp 44.

'Environmental Management Strategy for Colombo Urban Area', Draft Report published by the Urban Development Authority, Volume 1, March 1994, pp 12-24.

'Environmental Management Strategy for Colombo Urban Area', Draft Report published by the Urban Development Authority, Volume 111, March 1994, pp 21-24,75-79, pp 85-90.

Fernando M.S., 'Greater Colombo Flood Control and Environment Improvement Project', IESL Newsletter, Newsletter published by the Institution of Engineers, Sri Lanka, May 1994 (To be published).

Greater Colombo Wastewater and Sanitation Master Plan - Executive Summary, Prepared by Engineering - Science, Inc USA, April 1993, pp 12-19.

'Guidelines for Potable water supply, waste water and storm water drainage, domestic sewerage disposal and solid waste management of designated low income housing projects in Sri Lanka', Institution for Construction Training and Development, November 1991.

Inter-Agency Committee Report for the control of Vector and Nuisance Mosquitos, Published by the Central Environmental Authority, 1987.

Jayaratne K.A., 'Non-technical issues in the provision of sanitation for low income housing in Sri Lanka', Proceedings of the National Seminar / Workshop on sanitation infrastructure in low income hosing projects, National Housing Development Authority, December 1989, pp 28-36.

Morris D., 'Thinking things through', Water, Sanitation, Environment and Development, Proceedings of the 19 th WEDC Conference, Accra, Ghana, September 1993, pp 8-11.

Niedrum S., 'The need for hygiene education', Water, Sanitation, Environment and Development, Proceedings of the 19th WEDC Conference, Accra, Ghana, September 1993, pp 15-17.

'Study of the Canal and Drainage System in Colombo', Sri Lanka Water Supply and Sanitation Rehabilitation Project, Final Report, Volume 2, W.S. Atkins International in association with GKW Consult and RDC Limited, December 1988, Chapter 6.

Affordable sanitation for low-income communities

John Pickford, WEDC, UK

WITH HALF THE world's population lacking adequate sanitation, WHO's goal of *health for all by 2000* is unlikely to be achieved. A major difficulty, as in much other development, is shortage of funds. A postal survey conducted by WEDC in 1992 showed quite clearly that the most common reason for people not having latrines is that they cannot afford the cost of the types of sanitation being advocated in programmes and projects.

For many years *affordability* has been a major theme of international endeavour in our sector. Efforts have often been made to restrict the cost of providing water and sanitation to five per cent of average income, or the cost of sanitation alone to two or three per cent of income. In fact, many people pay much more than this when they perceive the value of what they get, or have no alternative. For example, the World Bank noted that in Onitsha, Nigeria, water cost slum-dwellers 18 per cent of the household income (World Bank, 1992). So it is now realized that *willingness to pay* is as crucial as affordability. The two must be considered together. What people say they can afford is usually what they are willing to pay.

Without outside funding, no expenditure on sanitation can be afforded by those with no income and by those at lowest sub-subsistence levels. They have to resort to open defecation, otherwise known as 'free-ranging' (Figure 1).

While this practice may be satisfactory for scattered rural communities, health hazards and difficulty in finding private places make it unsuitable for communities. Affordable improvement can be achieved by digging a hole and covering the excreta, as Moses commanded the children of Israel in the Sinai desert.

In some places low-income villagers set aside 'defecation fields' for open defecation. This leads to the danger of spread of hookworm unless sandals or other footwear can be afforded. Hookworm transmission can be reduced, at no cost except for labour, by forming ridges and furrows. People defecate in the furrows and walk on the ridges .

In urban and peri-urban areas various forms of 'dry latrines' have often been considered as low cost. In terms of construction alone the expenditure is undoubtedly low, as little more than a shelter is required. A receptacle of some sort completes the initial cost. Cheap containers for faeces are common. In India baskets were once usual, and I have seen old car or lorry battery cases, paint tins and discarded cooking oil tins employed. However, when the true total cost of a dry system is calculated it turns out to be an expensive option. Regular emptying of the container involves time, either by a paid scavenger or by a member of the household. The system is now universally condemned, but for many millions of people in many countries it remains in use as an affordable system

A somewhat similar system is 'wrap and carry', which from the users' point of view costs little or nothing. It is practised in many places worldwide. Defecation is onto a leaf, paper or plastic sheet, which is wrapped and dumped on vacant land or a refuse tip. A book published in the United States (Meyer, 1989) recommends wrap and carry for people enjoying the open air, it is *not* suitable for developing countries, even though its cost may be low. An exception is wrapping infants' faeces and putting them in latrines.

Pit latrines

By far the most common sanitation system in developing countries is one form or other of pit latrine. Pit latrines can be low cost, but many donors and other agencies make designs that are too expensive for low income people. Much of this paper is concerned with selecting designs, materials and methods of construction to make pit latrines affordable.

The basic purpose of a pit latrine is to concentrate excreta in one place, a hole in the ground, rather than depositing it indiscriminately.

In the pit faeces decompose, gradually forming a residual humus-like material that has no smell and is free from the pathogens that transmit diseases such as diarrhoea and worms (Figure 3).

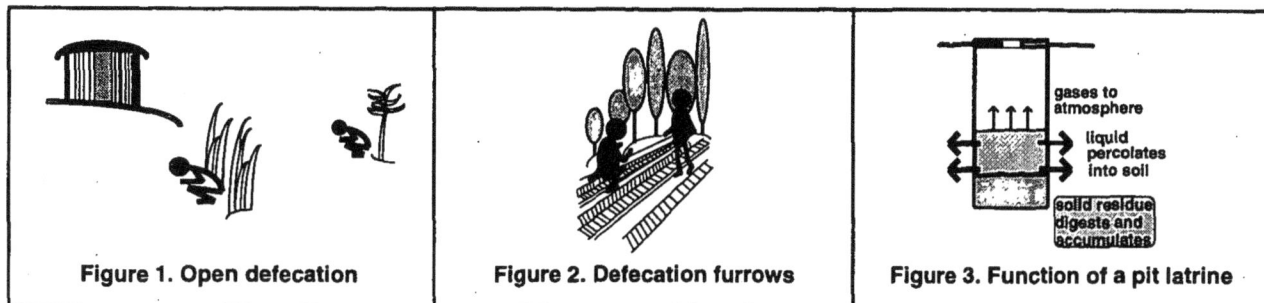

| Figure 1. Open defecation | Figure 2. Defecation furrows | Figure 3. Function of a pit latrine |

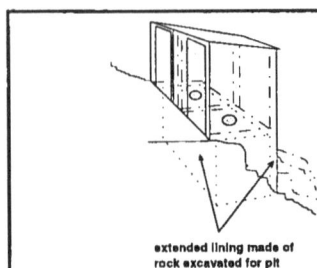

Figure 4. Latrine at George Brook

extended lining made of
rock excavated for pit

Figure 5. A basic latrine

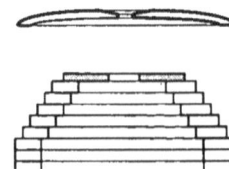

Figure 6. Domed slabs

However, in a pit which remains in use the decomposed humus is covered by fresh excreta that may be malodorous, may contain pathogens and may provide an ideal breeding place for flies. Dealing with these three problems (smell, germs and flies) are the fundamental issues that have to be considered to make a pit latrine sanitary and satisfactory for users. Users also require privacy, so some form of shelter is required and the nature of the shelter drastically affects the cost (and hence the affordability) of a latrine.

The pit itself

In relation to the *whole life cost* of a latrine, the largest possible pit is usually cheapest. A single small pit is initially cheaper than a large pit, but will not last long. Dividing the construction cost by the years during which the latrine can be used, the annual cost of a small pit may be high. The total annual cost per household (TACH) of large pits is likely to be lower. If the soil in which a pit is dug is stable when wet and dry, the size can be large enough to last for many years. I have looked down large pits that have used by African families for more than twenty years. The accumulated excreta was still two or three metres below the top.

Pits in unstable soil must be lined to prevent collapse of the sides. 'Standard' designs by external agencies often show these made of bricks, concrete rings or mass concrete, but in many locations cheaper locally-available material is used. For example in 1993 for the more-or-less standard twin pit pour-flush latrine in India the government and agencies like UNICEF paid householders 2400 rupees. In a village near Mysore latrines with the same design were built for 750 rupees. The saving was due to lining pits with stone obtained during well-digging and using lime mixed with a little cement for mortar in the shelter (Paramasivan, 1993).

Excavation is difficult where a pit latrine is built on rock or boulders. Raising the lining and floor increases the volume available for storage of solids. A compensation is the use of excavated rock or boulders for the lining, reducing the cost. Recently I saw an example of this in Freetown on a steeply-sloping hillside (Figure 4).

The floor slab

Crude latrines are found in some places. A couple of boards or logs are placed across the pit for users to put their feet when defecating. This leaves the excreta ex-

posed, with resultant smell and fly nuisance and chance of spread of disease (Figure 5).

A floor slab with a squat hole overcomes these problems (Figure 6). Where termite-resistant timber is available, an inexpensive floor can be made of logs, usually covered with a layer of gravel or mud. In many areas local craftsmen have developed techniques for making smooth hard mud floors which can be kept clean. Advantage should be taken of these skills, particularly where cement is expensive or difficult to obtain. Low cost improvements can be made with a thin 'skim' of mortar using a cooking oil tin-full of cement. SanPlat slabs can also be used' as discussed below.

Reinforced concrete (RC) slabs are normal for sanitation programmes where low cost is not a major issue. Costs become high for large diameter pits and we have already seen that in the long run large pits achieve savings. Three ways of providing cheaper concrete slabs for large pits are domed slabs, corbelling and enlarged excavation below the topsoil.

Domed slabs, as developed in Mozambique, need no steel reinforcement. They are thinner, lighter and cheaper than normal RC slabs and can be made by relatively unskilled people.

Corbelling with blocks, bricks or rock is suitable for linings that are circular in plan. A saving in the cost of RC slabs was obtained in low income areas of Karachi where fourteen sandcrete blocks (one part of cement to eighteen parts of sand) were used for the main lining and the corbelling reduced rings to seven blocks at the top.

Occasionally firm soil (or soft rock) is suitable for a large cavern-like excavation. A cost-saving small slab can be used with a lined shaft through soft soil near the ground surface (Figure 7).

Overcoming smell, flies and disease

Many users of crude pit latrines complain about bad odours and fly nuisance. Flies feeding on faeces are responsible for much transmission of disease. Flies, smells and health hazards are also, of course, major reasons for replacing indiscriminate defecation and dry latrines, and are associated with unsanitary emptying of full pits.

Three methods are commonly used for preventing nuisance from flies and smells in pit latrines. These are water seals between pits and latrine shelters, using tight-fitting lids and ventilating the pit in VIP latrines.

Water seals are the most effective of the three and are the first choice wherever water is used for anal cleaning and sufficient water for flushing is available. Lowest in initial cost is a slab and trap over a single pit. This system has been widely adopted in Bangladesh, where concrete rings are usual for the pit lining. Minimum cost, and hence maximum affordability, is for two rings. However, shallow two-ring pits have a short 'life', so may not be least cost in the long run..

Twin pits used alternately have become more-or-less standard 'best practice' in India. From technical and health points of view they are excellent and give best whole life value. Considerable subsidies were available in the past, but failed to benefit the poorest people.

In Medipur in West Bengal ten alternatives were offered at prices ranging from US$ 10 to $100.Ò All were pour-flush pit latrines. Apart from the two on the left, all can be upgraded by the householder, either by building a shelter or by constructing a second pit (Figure 8).

Past experience of wooden lids for squat holes has not been good, even in the United States (Wagner & Lanoix, 1958). However, the introduction of SanPlats in southern Africa has proved that tight-fitting concrete lids can be effective in controlling flies and smell. The secret lies in casting each lid in its own squat hole. Thin 600 mm square SanPlats reinforced with chicken wire can be made locally for a few dollars and are therefore generally affordable by low income communities. They only weigh about 35 kilograms (less than headloads carried every day by women) and can be fitted over traditional pole and mud floors. Where termite-proof wood is unobtainable SanPlats are made the same size as RC slabs. In addition to controlling flies and smells, they provide an easily-cleaned surface near the squat hole.

The ability of Ventilated Improved Pit (VIP) latrines to control flies was proved twenty years ago (Morgan, 1977).

Since then thousands have given satisfactory service wherever solid material is used for anal cleaning. Some have rectangular shelters with doors, others are spiral in plan without doors (Figure 9).

Many VIP latrines are fine structures, of which the owners are very proud. Foreign donors are often anxious to give 'the best' to the people they are trying to help. Consequently it is not unusual for a project to provide a few dozen very good VIPs, each costing several hundred dollars. If the funds had been used to help householders build their own low-cost varieties the overall benefit would have been infinitely greater.

In Zimbabwe Peter Morgan has developed a range of VIP latrines in addition to the well-known spiral type. He has rectangular VIPs using one bag, two bags, three bags and four bags of cement, with corresponding reduced cost. The one bag VIP has sun-dried brick walls and a thatch roof.

Ingenious variations have been introduced elsewhere, such as the type shown in Figure 10. This was built in Tanzania entirely of 'bush sticks', mud, cow dung and thatch. A particular feature is ending the spiral wall at the vent pipe, which has an effective locally-made fly-proofing at the top (Mugenyi, 1993).

Alternating pits

Building latrines with small pits seems an obvious way to make them affordable. The trouble is that they only last a short time before becoming full. I recently saw the folly of this practice in Freetown, Sierra Leone, where the many householders I spoke to spent an average of twenty dollars a year to have their pits emptied. The method of emptying there is similar to that common in West and East Africa, Myanmar and elsewhere. Solids removed from the pit are dumped elsewhere on the plot or nearby,

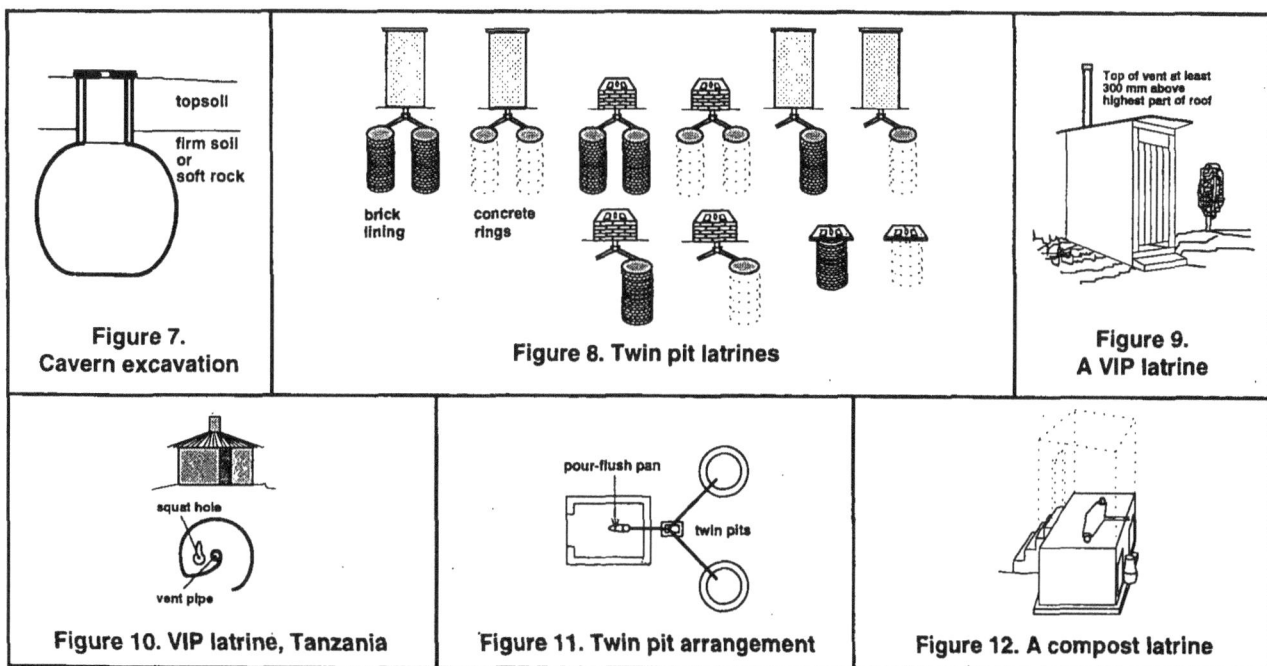

Figure 7.
Cavern excavation

Figure 8. Twin pit latrines

Figure 9.
A VIP latrine

Figure 10. VIP latrine, Tanzania

Figure 11. Twin pit arrangement

Figure 12. A compost latrine

with or without a thin covering of soil. Because the solids include recently deposited faeces the practice is unpleasant, and may be malodorous, fly-ridden and a serious health hazard, especially where worm infection is prevalent.

Building twin or double pits is cost effective, sanitary and is a valid alternative to large long-life pits. Each of the pits (or each chamber of a double pit) is only large enough for two or three years' accumulation of solids (Figure 11).

Compost latrines

Compost latrines, like the Multrum, are high cost and not affordable by low income people. Batch types, as shown in Figure 12, have been successful in Vietnam and Guatemala, but are only appropriate where there is a positive demand for compost.

The latrine shelter

Almost everywhere where there is a demand for latrines the main reason is not health benefit (or its converse a reduction of disease) but convenience and privacy. Convenience is best ensured by providing each household with its own latrine, although this is rarely possible for multi-occupancy buildings such as apartment blocks.

A screen made, for example, of bamboo and grass mats, provides privacy, which is especially important for women - there are many accounts of the distress experienced when women without latrines have to hold themselves until after dark. For UNICEF's programme in Bangladesh it was claimed that an affordable home-made bamboo shelter over a pit latrine should be the backbone of the sanitary revolution. In Botswana some concrete floor slabs were made with holes into which upright poles could be inserted by householders to make a simple shelter (Wilson, 1983).

However, in the early 1980s Bangladeshis were asked what they thought of latrines that were provided free. Most householders said the quality of the shelter was more important than the type of technology. Latrines were used more, especially by women, if the shelter was good (Gibbs, 1984).

Adding a roof provides protection from rain, sun and wind. In Srinagar many householders built shelters without roofs for pour-flush latrines. Wind-blown debris, leaves, twigs and the like caused malfunctioning (Sarma & Jansen, 1989)..

Because shelters are visible they provide status and a good shelter is often highly prized. This is fair enough where owners can afford to pay. Outside agencies also want status, so provide a few fine shelters which cannot be replicated by local people. It is not unusual to see blockwork latrine shelters well plastered and painted in villages where all dwellings are mud-walled and thatch roofed.

The following reasons have been given for the resistance of agencies to use appropriate cost-effective methods which lead to affordable sanitation (Amos, 1993):

○ they are unwilling to adopt standards that are inferior to those in developed countries;

○ professionals are reluctant to prepare schemes they regard as inferior to best practice;

○ external funding agencies often insist on standards which they consider will protect their investment;

○ innovative schemes require substantial research and design investment and have more risk than conventional designs.

Engineers, bureaucrats and politicians of national and local governments are often equally unwilling to adopt appropriate affordable practices for the same reasons.

Perhaps professionals should appreciate that the 'best practice' for preparing schemes is that which benefits the greatest number of people because they are affordable.

With so many millions of low-income people in need of adequate sanitation it is absurd that considerations such as those listed above should stand in the way of achieving progress towards sanitation and health for all by 2000.

References

Amos Jim, 1993 Planning and managing urban services. In *Managing fast growing cities* (Ed. Devas & Rakodi). Longman Scientific & Technical, Harlow. Pages 132 - 152.

Brandberg Bjorn, 1991. The SanPlat system: lowest cost environmental sanitation. In *Infrastructure, environment, water and people*. Proc 17th WEDC Conference, Nairobi, 19 - 23 August. WEDC, Loughborough. Pages 193 - 196.

Gibbs Ken, 1984. Privacy and the pit privy: technology or technique. *Waterlines*, **3**, 1, July, 19 - 21.

Meyer Kathleen, 1989. *How to shit in the woods: an environmentally sound approach to a lost art*. Ten Speed Press, Berkeley.

Morgan Peter, 1977. The pit latrine — revived. *Central African J Medicine*, **23**, 1 - 4.

Mugenyi George, 1993. *WEDC coursework*

Paramasivan S., 1993. *WEDC coursework*.

Read Geoffrey H., 1980. Aspects of low cost sanitation in Africa. *Report of the International Seminar on low-cost techniques for disposal of human wastes in urban communities*. Calcutta, February. Annexure IXfc.

Sarma Sanjib and Jansen Marc, 1989. *Use and maintenance of low cost sanitation facilities study of Srinagar city, Jammu and Kashmir*. Human Settlement Management Institute, New Delhi.

Wagner E.G. and Lanoix J.N., 1958. *Excreta disposal for rural areas and small communities*. World Health Organization, Geneva.

Wilson James G., 1983. The implementation of urban and rural sanitation programmes in Botswana. In *Sanitation and water for development in Africa*. Proc 9th WEDC Conference, Harare, Zimbabwe, April 1983. WEDC, Loughborough. Pages 46 - 49.

World Bank, 1992. *World development report, 1992: development and the environment*. Oxford University Press, New York.

This paper is based on parts of the author's book *Appropriate sanitation for low-income people*, to be published by IT Publications, London.

Urban sanitation issues in Sri Lanka

H. Pinidiya and K.M. Minnatullah, Colombo, Sri Lanka

SRI LANKA HAS committed to the target of total coverage by water supply and sanitation by the year 2010 through its national programs. Although the target seems ambitious, programs on water supply development progressed well (70 per cent coverage as of 1994), sanitation however lagged behind (50 per cent coverage as of 1994). In the rural areas a number of sanitation projects are implemented by different agencies with donor assistance, however, the densely populated urban areas of the country continue to be neglected. Some attempts have been made in the recent past to improve urban sanitation; these projects' primarily supply driven in nature' proved to be deficient in delivering the service to target beneficiaries. Project planning in isolation from the beneficiaries caused serious consequences during implementation and in effective use and maintenance, due to a lack of **ownership** of the facilities. Experience gained by documenting such examples could be utilized in future for better designing urban low cost sanitation projects. This study has been conducted to document lessons learnt from urban sanitation projects in Sri Lanka. The study reveals that, during the planning and implementation of urban sanitation projects factors such as demand for improved sanitation, socio-economic aspects, appropriateness of the technology, viable operation and maintenance mechanism, participatory approach in developing ownership, and a viable cost sharing/recovery mechanism were either lacking or inappropriately designed. As a result although targets in physical terms were fairly reached, the system was not sustained for long. Results of the study further indicate that, for those sanitation improvements where the community took a leading role and participated in addressing the above factors, the service was sustained.

Background

Nearly 3.7 million people (21 per cent) live in the urban areas in Sri Lanka, of which an estimated half resides in the low income settlements. The infant mortality rate in the low income settlements is between 32 to 54 per 1 000 live births, compared to the national average of 19.4 per cent. Prevalence of diarrhea, worm and parasitic infections, mal-nutrition, over-crowding, inadequate infrastructure services and higher incidence of non-schooling and drop-outs are some of the critical characteristics of these low income urban settlements.

From planning to implementation and subsequently during operation and maintenance of the urban sanitation projects, various constraints hampered the progress and functioning of the system. With a view to ascertain factors which impede the progress of urban sanitation projects and to identify issues that contribute to the non-sustainability of the facility, the UNDP/World Bank Regional Water and Sanitation Group for South Asia (RWSG-SA) and the Government of Sri Lanka (GOSL) agreed to conduct a study; this paper discusses the findings of the study.

For the purpose of the study an urban sanitation project implemented with the assistance of an external support agency in four densely populated urban areas in Colombo City was selected. A government institution was responsible for the implementation of the project. The characteristics of the study areas are similar to the urban areas in Colombo City and the four selected settlements represent the cross-section of the densely populated low income settlements.

The project

The project had three different types of sanitation systems— double-pit water seal latrines, common latrines connected to septic tanks, and individual water seal latrines connected to shallow sewers with common septic tank. The location of the settlements with number of households, type of sanitation system and the construction costs per household are provided in Table 1.

Table 1					
Site No.	Location	Sanitation system	No. of families	Cost per family US$	Year of construction
1.	Fg. Road	Individual water seal double pit latrine	375	89	1988
2.	N-Pura-1	Common latrine systems connected to septic tanks	450	14	1987
3.	N-Pura-2	Individual water seal latrines connected to shallow sewer with common septic tanks	110	310	1991
4.	K-tissa	As above	110	237	1987

The study

The study focused on the following factors to determine the relationship between these factors and the outcome of the project on effective utilization, ownership, maintenance, and long term sustainability :

(a) Demand responsiveness;
(b) Implementation strategy;
(c) Appropriateness of the technology options;
(d) Extent of community involvement;
(e) Functional aspects;
(f) Maintenance status;
(g) Cost recovery arrangements; and
(h) Sustainability of facilities.

A number of different techniques, including household surveys' were adopted to conduct the study. Most of the design and implementation details were collected by a study team from the key sector institutions and directly at sites. The survey was limited to 10 per cent random sample families from each of the four locations and the information gathered was cross-checked with interviews as far as possible to avoid recording errors. The survey questionnaires included data on technology options, implementation techniques, operation and maintenance practice, social concerns, functioning and acceptance of the facilities, willingness to contribute and participate, and other practical problems.

The limited time and resources available for the study did not permit an extensive in-depth study on factors identified above. Due to scientific limitations of the sample size, it was not possible to draw inference from the case load. Certain constraints restricted the scope of the study in the areas of (a) changes that have affected the day to day sanitary practice of the people (b) local variations according to socio-economic and cultural factors. However, the above limitations did not have any major impact on the overall study and the objectives of the study have been well covered. To record the functional aspects of sanitation technologies and the sociological aspects, the investigators had to establish close rapport for interpersonal dialogues with the study beneficiaries. To fulfil this requirement a number of prior visits were made by the investigation team to the selected project sites to develop better relationship and to be well acquainted with the community prior to collection of field data.

The most important results revealed by the survey have been summarized under the following groups in order to present clearly the outcome of the study :

(a) Socio-economic status and willingness to participate

The residents in these four communities have slightly different socio-economic status (Table 2) These low income settlements represent various ethnic groups mainly Sinhalese, Tamils and Muslims with different religions. The table shows the types of employment in the four areas.

Table 2				
Category of Earning	Fg. Road	N-Pura 1	N-Pura 2	K-tissa
Small commercial earnings			15%	10%
Labour and minor employees		20%	35%	60%
Informal sector activities	70%	75%	50%	05%
No definite earning	30%	05%		25%

Income per household varies from zero to Rs. 3000 per month, and the uncertainty in monthly income has crucial impact on their social status. Demand/interest for improved sanitation varies with the socio-economic condition of each community. The study revealed different views in the four locations.

Greater interest was expressed by the people who had knowledge of sanitation facilities, gained through their exposure to the middle income class at their work places. The project provided latrine substructure without any cost recovery mechanism, the households were to construct the superstructure. Families with a permanent monthly income are prepared to pay for the construction of latrines, however, a general lack of hygiene awareness and insufficient income is a vital factor for their passive attitude to sanitation improvements. It was apparent that a cost recovery system could be easily adopted in most of the areas and this would be an issue only for a very small percentage of households with a very low income. This is also evident with housing loans that the residents (most houses constructed by the residents are through loans provided by the government along with their own savings) have already committed to pay back to the government, and these loans are recovered without much difficulties. Most of the premises have sufficient space to construct latrine. The residents without latrines in the study area use common latrines provided by the government agencies.

(b) Acceptability and adaptability of the technology

The study confirmed that the residents were not consulted at the planning and design stages nor during implementation. The effects of non-involvement of the users are multifarious, which have affected the functioning, maintenance and adaption of the technology. The beneficiaries were dissatisfied with the common latrines which were not maintained, and because of their perception that the system will be well maintained by the provider (government). While all those who cannot afford household latrines were happy to use the common latrines, the study indicated that merely providing such facility did not solve the problem, instead it has created another set of problems. Improper construction also created a negative impact in the community - the most common double pit latrine system introduced in one of the study areas had all four walls of the rectangular pits plastered and the bottom completely sealed leaving no room for soakage. As a result the pits have overflowed within a short period after the construction and the system were abandoned by the residents, although design detail indicates that one pit is capable of handling the volume of human excreta of a family with four to six members for a period of one year. In certain places, it was found that the pits were used as water storage tanks for their daily domestic consumption.

In two of the study areas shallow sewers are connected to a common septic tank and the effluent from the septic tank is discharged to a nearby waterway through a tightly closed gravel filter. Functional details of the filter and the quality of the effluent discharged to the water way are not monitored. Therefore the effluent quality and the performances of the filters are unknown. The system is practically designed to discharge the effluent into the already polluted waterways in the vicinity. The study revealed that the filters were not properly functioning as the filter media was clogged. The users had no problem with this failure as it did not impede the function of the system.

The cost for each of the options adopted in the study areas noted in Table 1 shows that the cost of individual water seal latrines connected to shallow sewer system is extremely high. The system may not be appropriate for these urban low income settlements unless further cost reduction is attempted.

(c) Implementation and community involvement

The sanitation systems under the study areas were implemented by the government authorities through its contractors, except the superstructure which was constructed by the community after completion of the substructure. The authority supervised the construction of the substructure without any community involvement. The study indicates that the beneficiaries were not approached to participate during implementation as a result they expressed unfamiliarity with the system. The study further revealed that, the willingness for participation by the community in implementation was very high. An important feature revealed by the study is that during the construction stage no one had observed and corrected some of the gross deficiencies such as plastering the walls of the pits.

(d) Operation and maintenance

At locations where the community was provided with double pit latrines, the operation and maintenance problem did not arise since the system was completely abandoned due to failure of the technology. Beneficiaries expressed their dissatisfaction with the system and had developed a negative perception of the technology. In general the implementing authority had no prior interagency planning for the operation and maintenance of the system. The implementing authority expected the local authority to maintain the common latrines and septic tanks with shallow sewer systems. The local authorities did not consider maintenance as their responsibility since the provision of sanitation facilities for the low income settlements is not under their purview, as a result the maintenance of the systems were neglected. In absence of their involvement and ownership, the community expected the assistance of government authorities or any non governmental organization (NGO) to maintain the sanitation system. They also indicated that they do not have any organized capacity for this type of work.

Satisfactory maintenance practice was observed in only one area where the local authority carried out maintenance with the assistance of a community based organization. This process has shown the affectivity of community involvement in maintenance. In another location the community maintained manholes and sewers of the shallow sewer system. For septic tanks and filters which were beyond their capacity to maintain the community requested the assistance of the local authority, with no positive response.

With little or no interest from the local authority for the maintenance of sanitation facilities, the system has deteriorated beyond operation and the septic tanks have overflowed with clogged filters. Users strongly complained against this unsatisfactory condition. The sanitation project in these settlements therefore aggravated the environmental hazards instead of mitigating them and has earned public displeasure. There is no monitoring system to check the quality of the effluent discharged from septic tank through filters to the nearby channel. At places where the unfiltered effluent was directly discharged into the channel, the sewer system functioned without any apparent problem with the clogged filters. Community had no response to the adverse process because of ignorance of the system and lack of awareness of the health aspects.

Conclusion

None of the low cost sanitation systems reviewed under this study have succeeded in demonstrating sustainable improvements in the sanitation status of the urban low income communities. On the contrary statistical figures will support a full coverage in these areas and thereby portray an erroneous picture of the real situation. The study identified various reasons for the failures at different stages, from the planning and implementation to end-use and maintenance of facilities. Key factors that have contributed to the failure are listed below:

○ Lack of a demand-responsive approach in problem identification and planning is the overall factor which has negatively influenced the performance of the project including choice of technology, implementation, usage, ownership, maintenance and ultimately the sustainability of the system. The project had no provisions for proper dialogue or direct links with the community.

○ Cost recovery mechanism was not planned and introduced for the service. Socio-economic aspects identified during the study indicated the willingness of community to pay for the sanitation service provided. The findings confirmed that a reasonable cost recovery system could be introduced for the settlements under consideration. Such a mechanism if well conceived and implemented, would have created greater beneficiary responsiveness for the facility.

○ An important factor governing the community acceptance of the facility is appropriateness of the technology. The failures of the technology due to erroneous modifications and poor construction have bewildered the beneficiaries.

○ Poor planning for operation and maintenance and absence of clear agreement or consensus between the government institutions and the community have created a serious void in management of the system. There is no clear delineation of responsibility for the operation and maintenance of the sanitation system. As a result the system rapidly became non-functional, contributing to further degradation of the immediate environment.

○ In those areas, where, the community by own initiatives have collaborated with the local government authority over maintenance, the system functioned properly. Lack of confidence of the government in the capacity of the community has created a cul-de-sac in exploring the feasibility and extent of their involvement.

○ Greater disputes have been observed at locations where the common latrines were installed. As the system failed community expectation of the public sector for maintenance of the common latrines was dampened.

○ Implementation and maintenance responsibilities of the low cost urban sanitation system are with several agencies without any well defined jurisdictions. Agency and government level co-ordination is essential to resolve this confusion in order to define clear roles and responsibilities at various levels.

○ Public awareness campaign and hygiene education programs should be linked up at the outset of the implementation of urban sanitation projects for demand generation and also for proper use and maintenance of the facilities.

Acknowledgements

The authors would like to extend their appreciation to the officers of the sector institutions, NGOs and the community members who provided invaluable assistance during the study and in preparation of this paper.

References

Department of Census and Statistics, 'Statistical Pocket Book of The Government of Sri Lanka', 1993.

National Water Supply and Drainage Board of Sri Lanka, 'NWSDB Corporate Plan ', 1991.

National Housing Development Authority 'Evaluation of Low Cost Sanitation Projects', Colombo ,1992.

Ministry of Housing and Construction, UNDP/World Bank Regional Water and Sanitation Program 'Sri Lanka Water And Sanitation Sector', 1992.

Ministry of Health and Women's Affairs, 'Issues on The population of Sri Lanka' 1992.

Macrophyte trenches for septic tank effluent

P.R. Thomas, La Trobe University and T. Kalaroopan, Rural City of Wodonga, Australia

IN LOW- AND MIDDLE-INCOME areas of developing countries and in rural areas of developed countries, it is usual to provide septic tanks and associated drainfields or soakage pits as excreta disposal systems for individual households as well as for small communities up to about 300 people. Although such treatment systems are efficient and economical, septic tanks often have a bad reputation because of either inadequacy in the design of the unit or more often because of the failure of the drainfield which follows it. Many situations exist where the odorous septic tank effluent either ponds on the ground or seeps into the nearby stormwater drain due to poor performance of the drainfield causing health risks to people. This can be attributed to one of the following:

- inappropriate soil condition;
- excessive organic/hydraulic loading; or
- faulty design.

To avoid pollution problems the design and construction of the effluent disposal system is as important as the main part of the septic tank system. Proper functioning of the drainfield depends on percolation of the effluent into the soil profile and adequate aeration of the bed. Several alternatives to the standard drainfield system have been developed because of the pollution problems encountered with the standard drainfields (AWRC, 1988) and one such alternative is the use of macrophyte trenches or constructed wetlands. It has been shown that constructed wetlands have the potential to provide a relatively low cost, technologically simple method of wastewater treatment (Finlayson, 1983; Scholes et al., 1986; Davies, 1988).

Macrophyte trenches

A 'macrophyte trench' can be defined as a wetland specifically constructed for the purpose of pollution control and waste management, at a location other than existing natural wetlands. There are two basic types of constructed wetlands, the free water surface wetland and the subsurface flow wetland, the latter being considered to have some advantages over the other. In the subsurface flow wetland the flow is maintained below the media surface and there is little risk of odours, insect vectors or public health problems. A subsurface flow wetland system can be a discharge type or a non-discharge type, however, for an average household septic tank a non-discharge type can be reliably designed and constructed because of the low effluent flows. For higher flows, a discharge system can be used with potential for effluent reuse.

Aquatic plant species for use in macrophyte trenches treating wastewater should generally be selected using the following criteria (Mitchell, 1978):

- rapid and relatively constant growth rate;
- ease of propagation;
- capacity for absorption of pollutants;
- tolerance of hyper-eutrophic conditions; and
- ease of harvesting and potential usefulness of harvested material.

Also, it is preferable to select from native plant species which grow locally in the area. Examples of aquatic macrophytes that have been used in the artificial wetland systems are:

Floating plants:
- *Eichhornia crassipes* (Water Hyacinth)
- *Spirodela* (Duckweed)
- *Salvinia molesta* (Salvinia)

Emergent plants:
- *Schoenoplectus validus* (Great Bulrush)
- *Juncus ingens* (Giant Rush)
- *Phragmites* (Common Reed)
- *Typha* spp. (Cumbungi or Cattail).

These plants are able to transfer oxygen into the bed, creating aerobic microzones around the plant roots and anaerobic zones away from them. As a result, aerobic and anaerobic bacteria will both carry out the breakdown of the organic matter and removal of nitrogen through nitrification and denitrification processes. Wetlands can significantly reduce biochemical oxygen demand (BOD_5), suspended solids (SS), nitrogen, pathogens and metals through their complex chemical/biological processes as well as some uptake by the vegetation. Phosphorus removal in many constructed wetland systems is not effective because the gravel media offer limited contact opportunities between the wastewater and the soil, and due to the short hydraulic retention times. Removal of BOD_5 in the wetlands can be approximated by the first-order plug flow kinetics but they do not have a strong relationship with the hydraulic retention times as well as with aspect ratio (length:width) (Reed and Brown, 1992).

Home septic tank systems

For individual houses, macrophyte trenches to treat septic tank effluent can be designed and constructed to satisfy owners' landscape requirements in regard to position, and decorative plants such as lilies, cannas or ferns which grow well in wet conditions can be used to create an aesthetic value. *Typha* and *Phragmites* spp. are not recommended for domestic installations because of the massive seasonal release of wind-blown seeds (Mitchell et al., 1990). To dispose 1 m^3/d of septic tank effluent, conventional drainfields require a surface area of 50-100 m^2 whereas macrophyte trenches need 24-50 m^2 of surface area with virtually no pollution problems.

For an average household a single cell macrophyte trench 0.6 m deep is adequate to treat the septic tank effluent. The bottom of the trench should be lined with an impervious layer to prevent seepage if there is no clay layer. The inlet zone with a buried perforated pipe should have 25-50 mm washed gravel for a length of about 1 m, and for the full depth of the bed. Similar size gravel can be used for the outlet zone as well. The gravel in the treatment zone should be clean with sizes up to 15 mm in diameter. Since a low aspect ratio of the macrophyte trench is a very important factor in the hydraulic design of the system, a value of 3:1 or less is recommended (Reed and Brown, 1992). The cost of a home septic tank macrophyte trench system can vary above or below that of the standard drainfields and will depend on conditions such as topography, soil type, and the cost of gravel media (which represents over 50 per cent of the cost of subsurface flow macrophyte trenches). The longitudinal section of a single cell subsurface flow macrophyte trench is shown in Figure 1.

Treatment of secondary effluent — pilot studies

The pilot size subsurface flow macrophyte trenches at Wodonga Sewage Treatment Facility in Australia consist of four cells each 27.0 m long x 3.6 m wide x 0.6 m deep containing emergent vegetation growing in 0.5 m deep gravel media. Part of the secondary treated sewage from the treatment facility is used as the inflow to each of the four trenches, three of them planted with either *Schoenoplectus validus*, *Juncus ingens* or both species of plants and the fourth serving as an unvegetated control trench. The trenches with *Schoenoplectus validus* and *Juncus ingens* contain 10 mm and 14 mm size gravel respectively as bed media while 20 mm size gravel is used for the other two trenches. Monitoring began in December 1993 and mean concentrations of BOD$_5$, suspended solids, nitrate (NO$_3$), total phosphorus (Total-P), chemical oxygen demand (COD) and ammonia (NH$_3$) obtained to date from the inflow and outflow are presented in Table 1. During this period, flow to the trenches was varied to give different hydraulic retention times with one day being the lowest value.

The quality of the secondary treated wastewater from the sewage treatment facility has been inferior because of the unusual acceptance of effluents from a meat works and from a pet food industry to the treatment facility during the monitoring period. This affected the performance of the macrophyte trenches which were less than six months old and in the developing stage.

In the vegetated trenches BOD$_5$ removal efficiencies averaged between 56-63 per cent whereas in the unvegetated control trench BOD$_5$ removal averaged 79 per cent. In general the trench with both plant species, *Schoenoplectus validus* and *Juncus ingens* offered better BOD$_5$ and COD removal than the trench with either *Schoenoplectus validus* or *Juncus ingens*. Since BOD$_5$ removal is enhanced under aerobic conditions, it is reasonable to assume that the superior efficiency obtained in the control trench was due to the presence of oxygen in the higher voids component of the 20 mm gravel media. The macrophyte trench (20 mm gravel media) with the mixture of two plant species had slightly elevated effluent BOD$_5$ levels contributed by some of the decaying vegetation.

Suspended solids removal averaged 88 per cent with *Juncus ingens* and 80 per cent with both *Schoenoplectus*

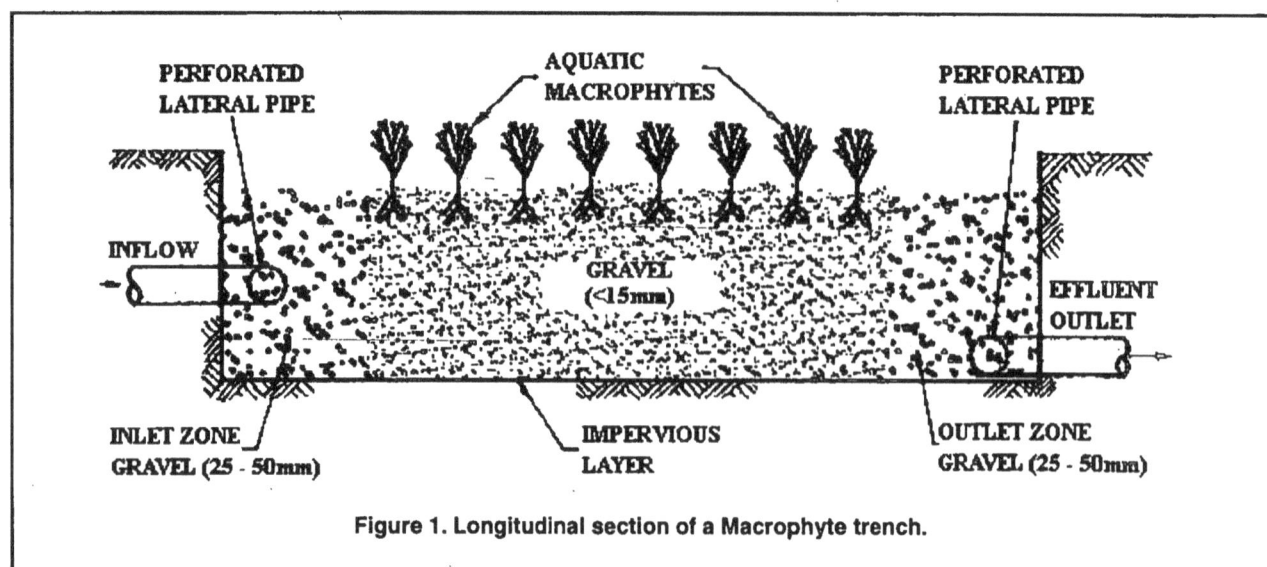

Figure 1. Longitudinal section of a Macrophyte trench.

Table 1. Performance of macrophyte trenches at Wodonga

Sampling period: December 1993 to April 1994	BOD mg/L	SS mg/L	COD mg/L	NH3 mg/L	NO3 mg/L	Total-P ms/L
Influent	57	50	221	29	26	8
Effluent from: Schoenoplectus validus (10mm gravel)	25	10	80	24	12	7
Juncus ingens (14mm gravel)	23	6	80	25	5	7
Mixed species (20mm gravel)	21	10	72	22	9	7
Unvegetated control (20mm gravel)	12	5	54	18	8	6

validus, and the combination of the two species whereas the control trench showed a reduction of 90 per cent. For suspended solids and BOD_5, results to date show that removal rates were not improved with long hydraulic retention times.

Reduction in ammonia was poor, 14-24 per cent for macrophyte trenches and 38 per cent for the control trench, whereas for nitrate removal, the overall performance was better with *Juncus ingens* offering the highest reduction of 81 per cent. Although it is expected to have a significant ammonia removal through nitrification-denitrification mechanisms from the macrophyte trenches, it is believed that the oxygen in the gravel media was insufficient to convert the ammonia to nitrate. Generally all of the trenches have been performing satisfactorily in regard to denitrification, with effluent nitrate levels dropping to as low as 3.5 mg/L in the trench with *Juncus ingens*. Phosphorus removal was low in all of the macrophyte trenches showing a removal efficiency of about 13 per cent. During this initial period of monitoring. surface flows were observed at the inlet end of the macrophyte trenches from time to time. This may be either due to excessive organic loading or due to the aspect ratio of 7.5:1 for each of the trenches.

Conclusions

Macrophyte trenches have the potential to provide an economically feasible and simple method of polishing pre-treated wastewater where a consistent high quality effluent is not always required. They can be integrated into septic tank systems for single houses as well as into the wastewater treatment systems for small communities, particularly in rural areas. Since the effluent from a septic tank normally has a BOD_5 of about 150 mg/L, either a two-cell septic tank or a septic tank effluent sand filter built in the system may be useful to reduce the loading on the macrophyte trenches to prevent any potential clogging problems.

References

Australian Water Resources Council (1988). 'Low cost sewerage options study'. *Water Management Series No 14*. Australian Government Publishing Service, Canberra.

Davies, T.H..(1988). 'Reed bed treatment of wastewaters: a European perspective'. *Water* 15(1), 32-33, 39.

Finlayson, C.M. (1983). 'Use of aquatic plants to treat wastewater in irrigation areas of Australia'. *Proceedings of AWWA 10th Federal Convention*, Sydney. Australian Water and Wastewater Association, 25.1-25.9.

Mitchell, D.S. (1978). 'The potential for wastewater treatment by aquatic plants in Australia'. *Water* 5(3), 15-17.

Mitchell, D.S., Breen, P.F. and Chick, A.J. (1990). 'Artificial wetlands for treating wastewaters from single households and small communities'. *Proceedings of the International Conference on the Use of Constructed Wetlands in Water Pollution Control*, Cambridge, U.K., September, 383-389.

Reed, S.C. and Brown, D.S. (1992). 'Constructed wetland design: the first generation'. *Water Environment Research* 64(6), 776-781.

Scholes, J.D., Kerr, R.I. and Nuttall, P.M. (1986). 'Treatment of wastewater by aquaculture systems'. *Australian Water Resources Council Research Project No. 80/13 Final Report*, Department of Resources and Energy, Canberra.

SECTION 4

SOLID WASTE MANAGEMENT

Rubber recycling

Rehan Ahmed and Arnold van de Klundert

GATHERING WASTE MATERIALS for recycling is least of all a new phenomenon as it done by tens of thousands of people in urban areas all over the world. Waste provides the poor people with a last resort to get employment through continuous struggle to survive with minimal income, bad working conditions and socially inferior status.

Enhancing the reuse of solid waste can restore some natural cycle and can contribute to solutions of urban issues like food production, waste disposal, energy shortages and improvement of environmental quality. Recycling decreases the quantity of waste to be collected and disposed of, provides job opportunities to the poor people, conserves finite resources and saves environment. The items commonly recycled are paper, glass, plastics, rubber etc. Recycling of rubber receives less priority and attention than other waste materials like paper and metals due to its financial value, margin of profit, final product, marketability, quality and public acceptance. This paper examines local technologies and legislative measures practised in industrialized and less industrialized countries and suggests actions for an optimal re-use of waste rubber.

Waste rubber production

Population growth, rapid industrialization and increase in living standards have caused extensive usage of rubber producing proportional quantities of wastes. Vehicular tires constitute the most important single item in terms of volume. In the industrialized countries rubber tires alone account for 60 per cent of the total rubber consumption. With the increase in automobile and bicycle production huge quantities of waste rubber tires are turning into waste mountains. In a modern car it is possible to find besides the tires more than 500 different rubber parts, whose total weight varies from 30 - 45 kilogrammes. In USA about 242 million tires are discarded annually. In Canada the figure is 10 million, while Germany wears out 0.6 million tires, France 0.4 million, U.K. 0.74 million and Italy 0.37 million a year. According to estimates one scrap tire per person per year is produced. In highly industrialized countries used tires constitute around 1 - 2 per cent of total municipal solid waste.

The remaining 40 per cent of the rubber waste mainly consists of tubes, conveyer belts, left overs from the shoe making industry. Other waste rubber produced consists of rubber parts, washers, insulation material in electrical appliances, packaging etc. and constitutes a smaller proportion.

Re-use, recycling and resource recovery technologies

In the industrialized countries tubes are mended only few times when punctured. Fairly good conditioned rubber tires are discarded within their life span due to over cautiousness, affordability, insurance, expensive manpower and high speed limit on highways. Damaged tires are seldomly repaired and re-used due to safety measures. Retreading and regrooving is not a common practice adopted, but is slowly gaining popularity due to stringent environmental laws, cost of tires and improved retreading technology.

According to fair estimates 40-50 per cent of the abandoned car tires and 60-80 per cent of the truck tires in industrialized countries are suitable for retreading and re-use. The recycling and re-use rate is fast declining due to failure of new technologies to penetrate the market as well as collapse of the reclaiming sector industries. Major quantities of rubber tires are stock piled and dumped on land. In the USA for example only seven percent of the waste tires are recycled, while eleven per cent are incinerated for their energy value.

The exhaustion of traditional disposal sites and strict environmental controls are contributing to a rapid increase in the cost of waste disposal services. In Germany and the Netherlands $ 75, in England and Denmark $ 46 and $ 26 per 1000 kg. Consequently, policy has shifted from a simple throwing away attitude towards prevention of waste generation and recovery of waste. Waste tires are made into crumbs and used for fuel. The rubber tires (32.5 MJ/kg) produce almost as much energy as liquid fuels (42 MJ/kg). This practice of obtaining tire derived fuel (TDF) is also fast declining due to availability of alternative fuels, environmental pollution and strict environmental regulations. Technologies are however available to keep the emissions within limits of the environmental legislation.

In industrialized countries creative and technologically sophisticated projects have been launched which utilizes huge quantities of rubber. The market for moulded rubber products is however growing due to higher process for virgin rubber and the public acceptance of products containing recycled rubber material.

The use of waste tires includes using it in artificial marine reefs, sport tracts, manufacturing of wheels for moving appliances and furniture, rubber flooring, protecting coasts from erosion, use in highway surfaces as additive to asphalt, drainage for sanitary landfill sites, insulation of foundations.

In the less industrialized countries waste rubber and discarded tires form a lower percentage of the total waste generated. Waste tires are re-used and recycled to the greatest possible extent: from large organized industries to unorganized and informal micro-enterprises. In these countries tires are regarded less as a waste problem. Local people find indigenous uses for them replacing other expensive and non available materials. Waste tires and tubes are segregated at household and communal level and collected by waste pickers who transport and sell them to dealers and small enterprises. Tubes when punctured are repaired by hot or cold method and used again more than its normal life. The tires are also repaired when cut. When the tyre looses its threads, retreading and regrooving is accomplished based on the available thickness of the tyre.

In these countries the recycling differs somewhat technologically due to type of tool, machinery, equipment the workshop owner can afford. Besides the craftsmanship is an important element. This causes a difference in quality and quantity of the recycled products and thus the market such products may be able to cover.

Waste rubber tires have different uses but only few have any big commercial potential. However hundreds of workshops or individual craftsmen process discarded tires into a variety of products. Basically there are three ways to recycle waste tires: direct product re-use, material re-use and energy recovery.

The first choice is repair and re-use of the product. All material of the product is saved. The least energy is wasted.

Examples of this category, whereby neither the original state is not changed nor any mechanical process is applied includes uses like making plant pots and holders, fenders for ships, harbour walls, crash barriers, fencing, the use of whole tires as walls for water wells, buffer blocks for traffic signs, oil spill containment booms, buffer for ship roads, covering material in agriculture and for children's sports.

In the second category the product is processed. It includes the making of footwear, containers, animal and hand carts, buckets, washbasins, doormats, ropes, harnesses and straps, flooring, parts for bicycles, furniture, pads, bushes, washers, gaskets, insulators, patching in tires, pipeline protection and other indigenous uses like rubber slates, mats, packaging etc.

The use of the rubber waste for energy use, is applied in the cement industries or in kilns for firing pottery or earthenware as well as to produce steam, electricity, steel etc. Discarded tires are also burned for another purpose: to recover the steel wires in the tread of the tire for packaging for example waste paper into bales.

Government policies and legislative measures

At present the government policies in the less industrialized countries are not much in favour of the recycling and reuse technologies. Some of these countries have regulations and legislation (to some extent) yet no incentives are provided by the government and municipal agencies to the informal recyclers. Practically government policies in the less industrialized countries are not geared in adopting the optimum recycling and re-use strategies. This is mostly due to the lack of awareness and giving low priority to the conservation of resources and utilization of waste. On the other hand there is bias towards high-tech modern solutions when equipment is bought. The funds spend on this by local governments and donors do however restrict the application of the equipment to well to do and central city areas. Moreover waste pickers are often hindered in the execution of their useful task in the removal of discarded waste.

If there is 'environmental' legislation, there is often a severe problem in the enforcement: there is either lack of budgets and trained personnel or companies dodge the regulations or bribe the executing officers.

In the industrialized countries the growing environmental awareness and the correspondingly growing legislation and the bureaucracy for the enforcement causes more and more waste materials to be recycled. Companies consider it part of their marketing strategy to show a 'green (environmentally sound) image' In the European Union (EU) the problem of the waste tires has gained attention. Agreements are made between industries and government on an approach to this problem. The first priority is to the decrease of the quantity of discarded tires through e.g. the reduction of the maximum speed on highways, a change in driving style and the extension of the life span of the tires.

The EU also suggests an increase in the distribution and sale of retreated tires in the private sector so as to reach a level of 25 per cent of the replacement market within a couple of years. Major retreading in Europe is done in Italy and Denmark. The retreading of passenger tires however has a bad image among car drivers and the difference in price between a new and a retreaded tyre is not big so far. Promotion of retreaded tires is therefore necessary.

Only 2-3 per cent of the discarded tires in the EU are ground to become granules. It appears that granulation could be boosted. The application of these granules is however limited since the granules can not be vulcanized again. Modern chemical technologies are however being developed to make these granules suitable for a wider application.

Figures on the recycling and re-use of waste tires adopted in less-industrialized countries show that for example in Mexico City and in Cairo, Egypt, 25 per cent and 22 per cent of the used tires respectively are being recycled. The recycling and re-use is mostly adopted by low income people to obtain a decent livelihood in a competitive market where human resources are abundant. Hundreds of people have gained useful employment in this sector now. Indirectly they are contributing towards waste minimization and resources conservation. Besides they provide a large number of low-income people with products they can afford. A market which is to open for the importation of discarded tires or governmental legislation restricting the collection of used tires, or the establishment of 'informal' workshops will seriously destroy such advantages.

To save the costs and foreign exchange, the government should levy import duties and regulations on virgin materials for promotion of recycling enterprises. The amount obtained can be spent on research and publicity of recycling. It is being noticed that with the import relaxation and minimizing the duty, the collection and utilizing of waste rubber has drastically reduced in Malaysia and Sri Lanka. Return of used items like rubber tires needs to be promoted by the producers/agents through refunding a small amount which will enhance better collection and utilization of waste tires.

Incentives

Recycling and re-use of waste rubber in less industrialized countries should be promoted. Incentives are to be provided to the waste recyclers and the people involved in waste collection and disposal. Regarding use of waste rubber utilization, the following incentives are recommended:

- official recognition of informal waste material trade by giving permits for waste collection, recycle and re-use to people and to allow less sophisticated (but safe) means of transportation into town;
- formation of small scale recycling unions and associations to be able to negotiate united with licensing departments and to exchange experiences;
- allocation of land on easy instalments at reduced cost to the recycling enterprises and at spots which can be reached easily by waste pickers and dealers/middle men transporting the waste materials;
- providing opportunities for on the job training and initiating short courses in easy local languages for low income people to be involved in recycling trade;
- involvement of NGOs, voluntary and civic organizations in recycling and re-use projects;
- disseminating information and education regarding providing health facilities and protective gears to the people in the recycling business, like masks, gloves and shoes;
- research and development of innovative, feasible technologies for the reprocessing of waste materials through small and micro-enterprises.

Waste rubber utilization strategy

Based on the detailed studies conducted by seven local consultants (5) the following strategy and sequence is recommended for optimal recycling of waste rubber tires:

- waste reduction at source
- repair and re-use (regrooving, retreading, casing)
- re-use as whole tires (artificial reefs, erosion control, crash barriers, fenders, thrust blocks, oil booms etc.)
- re-use after mechanical process into items of daily use (container, footwear, mats, flooring, parts of bicycle, etc.)
- recovering the raw material for making new products (natural or synthetic rubber, carbon blocks, fabric and steel wire);

- obtaining Tire Derived Fuel;
- shredding at land filling in mono fills.

Recommendations

On the basis of a literature study made on items and technologies adopted in less-industrialized and industrialized countries the following general conclusions are made:
1. In industrialized countries advanced technologies are adopted and usually supported by the government due to environmental considerations. A north-south co-operation is suggested like homogenizing the inputs and outputs in European Union and America. Economically viable and commercially feasible alternatives are to be adopted based on environmental legislation.
2. Based on the available tool, equipment, machinery and manpower, the low cost technologies are practised in the developing countries. There is a strong potential of utilizing the available local skills within a country and sharing knowledge within the developing countries.
3. The use of products produced from waste rubber are to be promoted. The entrepreneurs should be recognised and incentives to be given for increasing the quality and quantity. Appropriate legislation is to be framed, imposed and enforced favouring recycling, re-use and resource recovery.
4. Environmental effects are to be assessed and monitored for the recycling processes, products and working conditions.
5. Technical advisory services are to by provided by NGOs or institutions to the small scale recycling enterprises for improvement of recycling technologies.
6. Public education and awareness programmes are required for separating clean recyclable items and sending it to the recyclers. The products are also to be patronized and favoured.
7. The prime consideration of used rubber tires is reduction at source, without compromising the safety of vehicles. Tire life can be increased by modification of the tire construction, i.e. changing to radial technology and use of better weaving material.

References

[1] Bressi, Giorgio (1993), 'Recycling of used tires', ISWA Yearbook.
[2] Riggle, David (1992), 'New uses for old tires', *Biocycle* Magazine.
[3] Ryan, Felix (1982), *One dozen uses for old tires and tubes*, The Appropriate Technology Information Centre, India.
[4] Spencer, Robert (1992), 'New approaches in recycling tires', *Biocycle* Magazine.
[5] WASTE Consultants, Gouda, the Netherlands, WAREN Reports (1992): Six reports on the Recycling of ten Urban Solid Waste Materials (including rubber waste) made by local consultants in Cairo (EQI), Accra (AB & P), Manila (CAPS), Nairobi (USK), Calcutta (Ptr), Bamako (GERAD).

Solid waste management in India

A.P. Jain, G.B. Pant Institute of Himalayan Environment and Development, India

SOLID WASTE MANAGEMENT in India is an emerging and engaging area of study. However, the picture is often confusing and solutions fuzzy as information available in public domain is either scanty or scattered. This paper attempts to put together available information and analyse macro-tissues facing the Indian Techno-managers.

Though the core of the paper is built around urban solid wastes, a comparison with rural wastes is also provided. The comparison serves to highlight inherent strength of traditional knowledge systems in coping with rural wastes but also outlines the scope for modern S & T inputs. A separate section also outlines specificities of solid waste management in the Himalayan region. A series of conclusions and recommendations are reached after analysing the urban solid waste scenario.

Rural solid wastes

There are various types of wastes in rural areas namely community wastes, wastes from agricultural and agro-based industries, animal wastes and oil bearing seeds etc. Table 1 provides estimated annual generation of various types of rural wastes in India (Adapted from Report, 1990) (1).

The community wastes from rural area is estimated at five million tonnes of night soil and 10 million tonnes of refuse. The rural population of 629 million (Census, 1991) (2) is distributed over nearly half a million villages. This makes for a small population per rural settlement. Additionally, the population densities are also very low com-

pared to highly urbanized areas. Due to these reasons, collection and transportation of rural wastes in India is not a pressing problem. Low overall volumes also do not necessitate institutional structures for its management.

For using as fuel, animal dung is shaped into cakes and dried and stored to be used for domestic cooking. The excess is also used to make compost for farm applications. Composting is carried out by accumulating dung, domestic and other wastes in a heap or pit. Agricultural residues are largely used as animal feed; a small portion is also used as fuel and as construction material. Percent utilization of rural wastes for various end uses is outlined in Table 2 (Adapted from Report, 1990) (1). It is found that traditional practices of using wastes by way of fuel, animal feed and farm manure accounts for nearby 90 per cent of all waste utilization. Barely 1.6 per cent of wastes are not being utilized for any useful purpose. It is clear that traditional methods have been adequate in handling wastes generated.

Hence, rural solid wastes do not constitute a problem area like urban solid wastes. However, a case for S & T inputs does exist. Technologies such as Improved Chulhas (wood/dung stoves), Bio-gas from night soil or agro-residues, Biomass Densification, Gasification and Pyrolysis offer a way to enhance energy efficiencies and ensure efficient utilization of available resources, in addition to improving quality of life for rural masses.

The use of certain wastes as industrial raw material is limited to only 1.5 per cent. However, a number of possibilities exist to ensure increased industrial utilization of

Table I. Estimated annual generation of various rural wastes in India.

Types of Waste	Estimated Generation (million tonnes)	Percentage of total waste
Community	15	0.81
a) Night soil	5	
b) Refuse	10	
Agricultural residues	322	17.4
Animal dung	1365	73.74
Agro-industrial by-products	49	2.65
Oil seeds	100	5.4

Table 2. Percentage utilization of rural wastes for various end uses

End use	Percent Utilization
Fuel	14.8
Animal feed	35.0
Farm manure	39.3
Construction material	2.8
Industrial Raw material	1.5
Other uses	4.9
Waste	1.6

rural wastes for making value added products. Rice husk based particle boards, Bagasse based paper, charcoal, packaging material, chemicals through bio-conversion and industrial fuel are some examples.

Urban solid waste

With industrial progress, growing urban areas and resultant growth in urban solid wastes is a relatively new phenomenon in contemporary India. During mid-seventies, the per capita solid waste generation ranged from 150 - 350 gm/day for various Indian cities (Bhide et al. 1975) (3); whereas in late eighties, it ranged from 320 - 530 gm/day. The urban population is currently about one-quarter of the total population. It is projected to be nearly one third by end of the century. The total urban population in 2001 is estimated to be around 330 million from current 218 million. The class I towns alone account for nearly 60 per cent of all urban population (Census, 1991) (2).

The traditional knowledge systems primarily evolved for rural and dispersed populations have not coped well with densified living conditions and associated need for basic infrastructure and its management. Relatively poor management of urban wastes is reflected in degradation of living environment of urban areas.

Table 3 (Adapted from Report, 1989) (4) presents average per capita solid waste generation and collection efficiencies data for various categories of towns in India. Whereas the large towns show distinctly higher per capita waste generation; there is no significant difference between medium and small towns. The collection efficiencies in small and medium towns lag far behind at nearly 60 per cent compared to those of large towns at > 80 per cent. This points to weak infrastructure and poor financial status of small and medium towns.

Table 4 (Report, 1991) (5) presents physical characteristics of Indian urban garbage. The figures presented are on wet basis and the moisture levels can range from 40-70 per cent. The wastes largely comprise of bio-degradable organics. High moisture and organic content coupled with high prevailing temperatures make frequent removals necessary. This places additional burden on already over-strained system.

The collection of refuse presents peculiar problems as household wastes are thrown out indiscriminately. Also due to narrow lanes, collection vehicles can reach only selected accessible points. Hence, unskilled labour is used to sweep streets and collect garbage. Though labour rates are cheap due to large scale manpower deployment and low productivity, the costs are high. It is estimated that India spends four times as much on sweeping as on refuse collection (Pickford, 1983) (6). Poor motivation of workers, inadequacy of supervisory and management skills at local government levels are other leading causes of low productivity. The problem needs attention at appropriate levels.

The cost of collection in India tends to be a very large part of overall solid waste budget. To cite an example, the city of Ahmedabad with three million population and 1260 tonnes of solid waste per day, spends 85.8 per cent of its budget on collection, 13.4 per cent on transportation and only 0.8 % on final disposal (Report, 1990) (7). The benefits from present level of expenditure can be enhanced by following better methods of collection, efficient transportation, appropriate technology induction, better management practices and motivation of workers.

The three R's of waste management namely Reduce, Recycle and Recover are oft-repeated phrases in Indian policy circles. However, what is lost sight of is that culturally there is no propensity to waste. Also, there is a thriving informal sector of recycling. This recycling is achieved through Kabaris the waste handlers, who go from door to door and collect used bottles, broken plastics, metals, waste paper etc. This material is then traded for manufacture of secondary products for which markets exist.

Scavengers foraging through wastes is an unhygienic practice. It still however, contributes to recycling effort in India. The scavengers act as the second filter after Kabaris have taken away first batch of useful materials for secondary market. The income from foraging provides much needed subsistence to poorest of the urban poor. A ban on scavenging on health grounds may seem like a solution but will only aggravate the problem of basic sustenance. However, providing facilities such as bathing at the end

Table 3. Average per capita urban solid waste generation and collection efficiencies in Indian towns

Category of Town (Nos.)	Total Population	Waste Generated (Tonnes)	Per Capita Waste Gen. (gm/day)	Collection Efficiency (%)
Large (7)	22,312,961	11,761	527	82.8
Medium (17)	9,567,133	3,025	316	59.0
Small (9)	424,223	148	349	59.5

Table 4. Physical characteristics of Indian urban garbage.

Constituents	City-wise percentage of physical constituents				
	Delhi	Madras	Calcutta	Bangalore	Bombay
Paper & Card	5.88	5.90	0.14	1.50	3.20
Metals	0.59	0.70	0.66	0.10	0.13
Glass	0.31	-	0.24	0.20	0.52
Textiles	3.56	7.07	0.28	3.10	3.26
Plastics and Leather and Rubber	1.46	-	1.54	0.90	-
Wood, Hay and Straw	0.42	-	-	0.20	17.57
Bones etc.	1.14	-	0.42	0.10	0.50
Stones etc.	5.98	13.74	16.56	6.90	-
Fine earth and Ash, etc.	22.95	16.35	33.58	12.00	15.45
Fermentable	57.71	56.24	46.58	76.00	59.37

of day's work to foragers will go a long way towards improving their health status.

This market driven mechanism of segregation at source is a positive feature of Indian urban waste scenario. There are no figures available to estimate the volume of waste being processed, as this sector is not documented. The secondary product market though strong is not regulated. The specifications and quality is often sacrificed e.g. a lot of mixed plastic waste is reprocessed to make containers etc. These being cheap are bought readily by the consumers. However, storage of food stuff in these containers can be harmful for human health. There is a definite need to examine secondary products and to regulate effectively to safeguard human health.

Landfilling

Like other Asian countries, in India too most of the waste is landfilled. The methods followed are not in keeping with modern practices of sanitary landfilling. The wastes are largely dumped. This dumping is normally carried out in low lying areas which are prone to flooding. During rainy season, possibility of surface water contamination increases due to flooding of these low lying areas. The ground water pollution though largely unassessed is another threat posed by dumping of wastes. The daily cover techniques are poor leading to vector problems. The birds foraging on garbage dumps are known to cause substantial problems for aircrafts operating in the urban areas. The bird strikes have resulted in a great deal of loss to aviation sector.

This state of affairs results from lack of knowledge and skills on part of local authorities. Diversion of large part of money to collection and transportation of wastes results in non availability of funds for disposal activities. This forces local authorities to curtail even known precautions and practices and use short cut approach.

Composting

Composting is a highly suitable option for urban solid wastes in India. High organic content and moisture make it particularly attractive. Conceptually, the idea of composting is appealing as it helps to recycle the nutrients back to land. The process, however, requires segregation of inert material; which is achieved easily due to recycling by Kabaris and scavengers. This option, hence, appeared ideal in mid-seventies when a number of compost plants were set up in various cities. Mechanized aerobic composting offered hope for big towns starved of landfill space. Details of these plants are provided in Table 5 (Report, 1990) (7). The plants were commissioned during 1977-80 and were operated either by state agro-industries corporations or by municipal corporations.

These plants were expected to provide much awaited answer to growing problem of urban solid wastes but operational and other problems began to appear. Due to low skill/managerial inputs the operating efficiencies were low resulting in high cost of production. The problem was further compounded due to large distances between compost production centres and the compost utilisation centres, namely the farmlands. The resulting cost of transportation made marketing even more uneconomical. Farmers also reported problems with broken glass pieces in the compost.

Table 5. Details of pilot compost plants

S.N.	City	Capacity (tonnes/day)	Operated by
1.	Bangalore	200	Karnataka Compost Dev. Corp.
2.	Baroda	150	Baroda Municipal Corp.
3.	Bombay	300	Bombay Organic Manures Ltd.
4.	Calcutta	300	Calcutta Municipal Corp.
5.	Delhi	150	Municipal Corp. of Delhi.
6.	Jaipur	200	Rajasthan State Agro-Ind. Corp. Ltd.
7.	Kanpur	350	Kanpur Municipal Corp.

The composting, however, still remains a strong option for small and medium towns. Semi-mechanized aerobic composting is ideally suited to waste volumes in these towns. It demands less in terms of operational and management skills. The product off-take can be good due to close proximity of agricultural areas to almost all such towns in India. The problem of broken glass can be taken care of by suitable local legislation to ensure segregation at source.

Incineration

Incineration is not a total solution for solid wastes. The inert remains still have to be landfilled or used otherwise. This acts as a volume reduction step. In India, it has not found much use as the garbage tends to be low in calorific value and volumes are generally low for a central facility. The technology for incineration is not available indigenously and import options are highly capital intensive.

During 1980s an incineration plant was set up at N. Delhi at a cost of Rs. 220 million or US$ 6.9 million (May 94). This 300 TPD plant was set up using Danish technology with assistance from Danida. It was also expected to generate power for local grid. The operational experience was not satisfactory. The desired calorific value garbage did not reach the facility as a result of prior segregation due to market mechanisms and scavengers.

Despite apparent failure of this attempt, incineration will remain an option for future and experience gained in this venture will be useful. In the meanwhile, incineration on smaller scale with or without energy recovery will continue to be a viable option in a number of location and waste specific cases such as hospital wastes.

Anaerobic digestion

For high moisture and organic content of Indian wastes, the anaerobic digestion is another suitable option. However, there are no ready technologies available for processing heterogeneous material such as urban solid wastes. The existing methods are suited to homogeneous materials. The costs of cleaning and separating mixed heterogeneous wastes are likely to be high.

A good way to avoid these problems is to intercept suitable wastes at the point of generation before it is mixed with other wastes. Kitchen and vegetable market wastes are largely suited for this purpose. These wastes can be collected and treated at source, if space permits. The resulting bio-gas can be used for captive energy use such as lighting and cooking etc.

Few bio-gas systems are currently available to treat wastes of fruit and vegetable origin (Nagori et al. 1988) (8). Though currently unfeasible as a large scale option, bio-gas systems can effectively handle localised and specific wastes and contribute to environment friendly disposal of wastes.

Refuse derived fuel (RDF)

This method of waste disposal primarily views waste as a resource. After separation and size reduction, the combustibles can be pelletized. Integrated Waste Management project at Bombay attempted to do just that. Due to local conditions, the product off-take and price realization was estimated to be good. This avoided the earlier problem faced by composting plants. The large scale processing of garbage was also supposed to slow down exhaustion of landfill space considerably in the near vicinity of the city; obviating need to spend much larger amounts on transportation costs.

This pilot technology development effort also offered prospect of totally indigenous and cheap technology. The cost of 80 TPD plant was Rs 15 million or nearly half million US dollars (May 94). This compared very favourably to N. Delhi incineration plant (300 TPD, Rs 220 million). As it was first attempt of its kind, it required experimentation and modifications to zero down on specific waste handling, size reduction and separation processes along with optimization of system parameters.

The plant was erected and extended trials were undertaken. A number of new innovations were made in garbage separation methods. The fuel pellets produced were also test marketed successfully. However, there was a need to support the technology development effort for a long enough duration which has been lacking.

Despite the promise of RDF, it will be limited in application due to need to have large industrial areas in close proximity to market the fuel to. The cost differential between cost of coal and the RDF should also be attractive to ensure sales.

Solid waste management in the Himalayan region

The Himalayan region of India is spread across 12 states and accounts for 18 per cent of country's land area and six per cent of the population (Swarup et al. 1994) (9). The region is largely remote and comprises of far flung and difficult to access settlements. The population is largely rural. The urban areas comprise small and medium towns. The region also receives a good number of tourists. Proper management of solid wastes is of paramount importance in this region because of its increased pollution potential resulting from down stream effects.

Rural population relies on surrounding forests for its energy needs. This has been one of the causes of forest depletion. It is difficult to reach conventional urban energy sources such as bottled cooking gas, kerosene etc. to remote rural areas due to difficulties of access in mountain terrain. This makes role of non-conventional energy sources quite important. The rural wastes also assume significance due to their energy potential. The role of simple technologies to ensure efficient utilization of energy from waste will hence be very important in this region. The bio-gas option will have limited application due to lower prevailing temperatures in the region. It may however be feasible upto certain altitudes. Improved wood/biomass burning stoves and biomass gasification technology can play important role.

The urban waste disposal options will be considerably affected by mountain specificities. Landfilling may not be possible due to undulating terrain and paucity of flat spaces. Composting may be predominant choice but with due care to intercept run-off from composting areas and its treatment.

The seasonal flow of tourists accounts for a good deal of floating population especially during summer. Any planning for solid waste should consider this factor. The large influx of tourists has also resulted in problems of litter at high altitude scenic and tourist spots. This has created peculiar problems of waste retrieval and restoration of those areas.

General Conclusions

1. Bio-gas systems can effectively handle localized and specific wastes and contribute to environment friendly disposal of wastes.

2. The semi-mechanized aerobic composting, however, still remains a strong option for small and medium towns.

3. Mass incineration will remain a possibility for future.

4. Incineration on small scale will be indispensable for hospital waste etc.

5. RDF in specific cases is an attractive option provided that sustained indigenous technology development efforts are made.

6. All technologies attempting to process garbage are difficult to master as they are traditionally geared towards handling virgin and homogenous materials.

7. The role of simple technologies to ensure efficient utilization of energy from waste will hence be important in the Himalayan region.

8. Waste generated by floating population of tourists is an important consideration in the Himalayas.

9. Though a number of options are suited to Indian conditions, a particular solution should take into account location and waste specific factors.

10. No single technology option will be sufficient to take care of emerging problems of urban solid wastes. A mix of options will have to be developed and applied on case to case basis.

Recommendations

The recommendations can be broken up under various categories such as:

- Manpower/education and training
- Regulatory and fiscal
- R and D and technology development

Manpower/education and training

- Local governments should work towards infusion of greater management and supervisory skills.

- Data and knowledge base at the local government level should be strengthened.

- The problem of over-staffing should be given serious socio-political consideration.

- Education efforts should focus on women to highlight proper household disposal, segregation and community participation.

- Steps should be taken to improve health status of scavengers.

Regulatory and fiscal

- Local bodies should awaken to the need for suitable legislation as per the prevailing local conditions.

- Privatization of collection and transportation of urban solid wastes is highly recommended. It will help provide a cap on expenditure, reduce inefficiency and provide better level of service.

- Private initiatives in waste disposal or utilization should also be encouraged by way of fiscal and other incentives.

- A nominal garbage tax along the lines of house tax is recommended. It will generate much needed finances and also bring into focus much neglected problem of solid waste.

- Secondary products should be regulated to protect consumers.

R and D and technology development

o Standard practices for sanitary landfilling under Indian conditions should be developed.

o Technology and training packages for semi-mechanized aerobic composing at small and medium scale should be developed.

o R and D efforts should focus on developing improved plants for wastes from vegetable markets, kitchens and restaurants etc.

o Recycling in the informal sector should be quantified.

References

1. 'Rural Waste Management', A report by National Waste Management Council, Ministry of Environment & Forests, Govt. of India, 1990.

2. Census of India, 1991.

3. Bhide A.D. et al., 'Studies on Refuse in Indian Cities, Part II - Variation in Quality and Quality', *Indian Journal of Environmental Health*, Vol. 17, No.3, July 1975, P. 215-225.

4. Study on Delivery and Financing of Urban Services by Operation Research Group for Planning Commission of India, N. Delhi, 1989.

5. City Garbage Treatment, A position paper by Business Horizons for Technology Information Forecasting and Assessment Council, Dept. of Science and Technology, N. Delhi, India, 1991.

6. Pickford John, 'Solid Waste Problems of Poor People in Third World Cities' in *Practical Waste Management*, Ed. J.R. Holmes, John Wiley and Sons, 1983.

7. Report of Sub Group on Urban Municipal Waste Management, Ministry of Urban Development, Govt. of India, 1990.

8. Nagori G.P. and Rao C.S., 'Biogas Manure Plants Based on Agricultural Residues', SPRERI, Vallabh Vidyanagar, India, 1988.

9. Swarup R. et al., 'Land Holdings' Scenario in the Himalaya', in *Himalaya : Past and Present* Vol. III, Eds. Joshi M.P. et al., Shree Almora Book Depot, Almora, India, 1994.

Supporting and strengthening junk dealers and recyclers

Danilo G. Lapid, Centre for Advanced Philippine Studies, Philippines

IN 1973, the City Government of Manila passed an ordinance regulating scavenging in the city by requiring scavengers to operate in non-tourist areas and secure permits. Otherwise, they would be fined (P20.00 to P100.00) and/or imprisoned from one to six months.

Scavengers are an integral part of the junkshop-recycler network. These junk shops and small recyclers have come a long way as waste managers. Now, they are being recognized by the larger community as agents for urban environmental protection. There is a growing realization, at the national and local levels, that the alternative waste pathway through junkshops and recyclers is more beneficial for the common good of the public, the government and the environment than the usual waste collection and disposal system currently being managed by the government through dump trucks and landfills. There are about 1000 junk dealers and about 300 recyclers in Metro Manila.

This paper has three major topics, namely, 1) previous government projects dealing with junk shops and pushcart collectors; 2) efforts of NGOs and individuals in supporting and strengthening this informal sector; and 3) some legislative agenda for junk shops and recyclers.

Past experiences that formalize the recycling activities in Metro Manila

In July 1978, the Cash for Trash Program was presented to the Minister of Human Settlements by some concerned citizens. The plan was to implement a solid waste separation and resource recovery program. The original plan envisioned the use of the existing informal system of resource recovery as the cornerstone of the proposed plan. A budget of P 1.8 million (approximately US$200 000) was released to fund the program.

During the implementation, the plan was modified. Instead of using the existing informal system of resource recovery, the implementors under the office of the Deputy Minister of Human Settlements created a parallel system to directly compete with the informal system. Eco-aides were recruited as collectors. A single recycling corporation was established to buy all the recyclables from the 30 eco-centres (read junkshops) that were subsequently established.

Barely two years later, in 1980, all the eco-centres closed shop. Analysts were in agreement that the primary cause of failure was the adverse reaction of the existing informal system most notably the junkshop dealers to the project.

Implementors of the aborted project failed to give importance to the role of the informal system in the recycling of waste materials. As a result, the junkshop dealers felt threatened by the establishment of the project because they were left out of it. The dealers saw to it that the eco-aides sold their collection of waste materials to them by giving these eco-aides better prices than the buying prices of eco-centres. The eco-aides then returned to the eco-centres declaring that they failed to buy anything.

Some dealers that were interviewed also attributed the collapse of the program to the alleged gaffe committed in the procurement of fixed assets for the eco-centres. Eco-aides were provided with overpriced pushcarts, overpriced weighing scales, overpriced warehouses, etc. These purchases led to the depletion of the seed capital needed by the eco-aides to buy the waste materials.

Overall, the collapse of the project could be traced to the failure of the program to integrate or include the informal sector into its program hence the kind of reaction and competition that it had not bargained for.

Private initiatives by NGOs

In 1983, the Women's Balikatan Movement of the Philippines started organizing existing junkshop dealers in San Juan, Metro Manila using the original concept of the Cash for Trash program, that is, to buy recyclable items as part of their aim to protect the environment.

The Balikatan campaigned for waste separation at the household level, started to organize the junk shops in a Metro Manila municipality and link them with residential communities to make the collection of recyclables more efficient. It acted as guarantor on behalf of the junkshops so lending institutions would provide them some credit for working capital. It also popularized the term 'eco-aide' to refer to scavengers and pushcart collectors, thereby highlighting their role in ecological care.

At the start of the project, Balikatan got the co-operation of the municipal government of San Juan. Upon instruction from the Mayor, community assemblies were convened in each of the 21 barangays (local districts) of San Juan to explain the concept of solid waste separation and the benefits that it will give to the households. Discussions were held on how to increase solid waste recovery. Permission was obtained from the government water utility company for the use of one of its vacant lots as a junkyard or collection centre.

Now, Balikatan operates in other municipalities of Metro Manila. The group has started to organize the junkshops into co-operatives to avail themselves of government subsidies and credit and enable them to get better deals with bigger junkshops and recyclers and even formal institutions.

Through the intercession of Balikatan and other groups, other government support has started to pour in. The Department of Trade and Industry is in the process of approving a P250 000.00 (US$9 000) soft loan package for eco-aides to be used as working capital in their daily chores of buying recyclables from households. Recently, the Department of Social Welfare and Development field office in Region 3 released the P100 000-livelihood financial assistance to a group of eco-aides as seed capital for income-generating projects.

Other NGOs have joined in. The Ayala Foundation, Inc. wants to campaign for the change of name of junk dealers to 'waste managers'. Waste separation from the source is now becoming a popular campaign of many NGOs especially the Recycling Movement of the Philippines and the Ecology Centre through their community-based zero-waste campaigns. Business and civic organizations, like the Rotary Club, are also willing donors to community-based waste management drives.

Recently, the Centre for Advanced Philippine Studies (CAPS) launched a project creating a Waste Management Resource and Information Centre (WMC). The idea of a WMC arose two years ago when individuals, NGOs and home owners started inquiring from CAPS where to sell or what to do with various waste materials. Before that, CAPS conducted studies on recycling and urban environmental management through communities and NGOs. Apparently, CAPS then became known for waste management matters.

This centre will primarily serve as the main repository of data and information on solid waste management. It will have a computerized on-line inquiry system on junk shops and recyclers available to anybody. The second major activity is the Information and Education Campaign where waste segregation, and the role junk shops and recyclers will be promoted. Another major component is the training of community leaders and organizers on solid waste management. Although the centre will be managed by CAPS, it will be sensitive and will cater to the needs of the eco-aides, junkshops and recyclers, as well as other waste conscious NGOs and entities.

Maybe of interest to this conference is the question of affordability and feasibility or sustainability of such an information centre. The WMC got its initial funding from a corporate foundation. The grant was enough to launch the project and sustain the Centre for one year. Last April, we conducted a strategic planning workshop to plan for the years to come.

First on our list of 'Do's' is institution building. This means inviting personalities in the private and government sectors known for their commitment to environ-

mental concerns to join the Centre as officers and Board members and/or sponsors. Given their stature and influence, they can help generate donations, grants and/or endowment funds for the Centre. This also means employing dedicated and qualified personnel to run the program.

Second is the program itself. Our vision for the Centre is for it to become a technology clearing house, a training venue, and a research and development institute for solid waste management. Anytime anyone thinks of waste management, the WMC should come to mind. The Centre should be able to sustain itself through its services. WMC can enter into service contract agreements with city or municipal governments, corporation home owner associations and NGOs to help them formulate waste management plans and train their personnel. WMC can also raise funds through research contracts and subscriptions from its publications.

Last but not least is to obtain government support for the program. Recently, the information office of the national government endorsed our radio ad for free airing in 20 radio stations that has national coverage. We hope to enlist government support for other components of our information campaign through television and the print media. This is as good as approved because others, such as the Recycling Movement of the Philippines and the Green Forum Philippines, have already obtained government endorsement. Another venue of support is in publication. Under negotiation with the Quezon City Government is the printing of our 3-R's handbook.

Legislative agenda

As mentioned, the government has begun extending a helping hand to junkshops, eco-aides and NGOs in their various ways of curbing solid waste. The good news is that even the legislative branch of the government has joined the crusade.

There are now a total of 11 congressional bills, seven in the Senate and four in the Lower House, pertaining to waste segregation, reduction, recycling, handling and disposal of solid and liquid waste. The salient points of these bills which are pertinent to this paper are: (i) the formulation and implementation of an integrated (national and local) solid waste management program for waste minimization, (ii) inclusion of waste segregation, reduction, composting and recycling in the school curriculum, (iii) tax exemption for all anti-pollution and recycling devices, (iv) year-end awards to government agencies and private organizations for outstanding projects, (v) use of recyclable materials in the packaging of consumer products, (vi) prohibition of the manufacture and use of styrofoam food containers in all food establishments, and the (vii) regulation of the manufacture, distribution and use of non-biodegradable plastic bags.

We are still waiting for a bill that will regulate or ban the importation of waste materials for recycling since they depress the market for locally collected recyclables. The livelihood income of the eco-aides and junkshops are significantly lessened when there is a glut of such imported waste materials.

Conclusion

The role of junkshops, eco-aides and recyclers in environmental protection are now being positively recognized by the larger community, unlike the time they were looked down upon and threatened by government established competition. Because of their perseverance, they now represent an effective alternative to the dominant but expensive way of collection and disposal of waste. The NGOs and the government sector have come forward in their support through credit assistance, organizing, research and publication, media promotion and legislation. I sincerely hope the trend continues so they would become an integral part of the management of waste in the urban centres, that is, through legislative action on waste segregation and minimization, the eco-aides, junkshops and recyclers will soon become the dominant players in the management of solid waste. Through them, the income from waste collections and disposal will be democratized or decentralized from the few politically favoured operators.

Small-scale urban organic waste recovery

Inge Lardinois and Arnold van de Klundert, WASTE Consultants, Gouda, The Netherlands

ON REQUEST OF the Undugu Society of Kenya (USK) a comprehensive research was carried out (1991-1993) addressing both problems of unemployment and uncollected waste in low-income areas. Focussing on resource recovery, the research covered the great variety of products made, markets covered and technologies used by small-scale enterprises in six cities in the South. WASTE Consultants has co-ordinated this research project, whereas the research itself was carried out by consultants[1] in the following six cities: Manila, Calcutta, Cairo, Nairobi, Bamako and Accra.

This paper is based on their findings, on field visits and additional literature and deals specifically with the recovery of organic waste, which was one of the ten materials researched[2]. The paper will focus on small-scale recovery methods of urban organic waste and describes three examples from Bamako, Cairo and Manila. It will also deal with economic feasibility, environmental and health aspects, possible measures for improvement and the role of government authorities.

Background

In many cities in low- and middle-income countries, the municipal refuse collection and disposal services are woefully inadequate and thus, waste accumulates in the streets and at transfer stations. A large proportion of urban waste consists of organic material, which therefore forms an interesting source for recovery. It may offer ample opportunities not only for the improvement of environmental and health conditions but also for employment generation.

All over the world, municipal authorities have started initiatives in this area, the most important being composting. However, large-scale installations have been purchased that were often too expensive, too complicated, and not tailored to local conditions. The construction and operating costs of these highly mechanized facilities were often higher than the revenue received from the sale of compost. Consequently, some facilities have been closed down, others have been scaled down, and many operate well below their planned capacities. A tentative conclusion is that in the urban areas of economically less developed countries, large-scale reprocessing of organic waste is undergoing a crisis. Already existing and new small-scale recovery opportunities could therefore provide valuable ingredients for an alternative strategy.

Organic waste can be the basis for many diverse activities. Basically, organic material can be re-used in three ways: to improve the soil (compost), to feed animals (fodder) and to produce energy (briquettes, biogas). The direct re-use of organic waste as fuel for cooking in the home is quite common. Woody residues such as coconut shells are frequently used when conventional fuels are either too expensive or difficult to get hold of. To achieve a more efficient use of existing organic waste resources, experiments are carried out with briquetting, which involves the compressing of combustible materials into a solid fuel product that can be burned like wood or charcoal. Another way of converting waste into energy is anaerobic digestion whereby not only energy in the form of biogas is produced, but also compost. Experiments are currently under way to study the anaerobic digestion of municipal organic waste in reactor systems. However, at the moment these are not profitable activities. In terms of providing a source of income for small and micro-entrepreneurs, the first two opportunities (the use of organic waste as compost and as fodder) are already practised to a certain extent.

Economic feasibility

The recovery of organic waste differs from the reprocessing of other waste materials such as plastics and rubber in that the latter have proven their economic profitability in small-scale enterprises: in Asian cities hundreds of such micro-enterprises exist. In general, organic waste reprocessing does not take place at such an extended scale; only animal raising as income-generating activity is carried out by many entrepreneurs, although on a part-time basis. We will now give a few examples of feasible small-scale enterprises reprocessing organic waste:

Bamako: compost making

In Bamako, Mali, municipal solid waste is decomposing via natural processes. Micro-entrepreneurs exploit this free source of raw organic material. When the garbage is delivered to the dump, it is stacked into piles which are left to decompose. The process is not controlled, and the piles remain undisturbed for an indefinite period. Using relatively simple tools (sieves, spades, brooms, pushcarts) the entrepreneurs manage to produce fine compost by sieving out the impurities and non-organic materials. They earn about three times the minimum wage, also due to the absence of transportation costs. The resulting compost has a good appearance and is almost free of visible foreign matter. The regular supply, the low price and the proven quality of this soil conditioner have created a high demand, particularly from vegetable farmers in the peri-urban areas.

Manila: pig raising

One of the simplest ways to recover the value of organic waste material, is to feed it to animals. In the outskirts of Manila, the Philippines, pig raising is a popular backyard operation. Commercial animal feed is substituted with organic waste, which costs less than 50 per cent of commercial feed. Pig raisers collect the organic waste on a daily basis from restaurants in the city centre. The use of organic waste as pig feed reduces production costs considerably: it doubles the net profit per production cycle of 3.5 months. Given that pig raising is a part-time activity, the earnings per day are reasonable compared to the minimum wage level.

Cairo: pig raising and co-composting

The Zabbaleen, a marginalized group of Coptic Christians in the predominantly Islamic society of Cairo, survive through the resource recovery of various waste fractions. They earn their daily income by collecting relatively high-value waste from middle- and high-income areas of the city. Valuable materials, such as plastic and paper are sorted and reprocessed in a large number of micro-enterprises. The Zabbaleen also raise pigs on the organic material they find in the refuse in so-called *zeribas*, or enclosed courtyards. Once or twice a year the *zeribas* are cleaned and the mixture of leftovers and pig manure is carried on donkey carts to the composting plant. Normally, the plant processes 30 tonnes of compost per shift per day. During the season when land is prepared for cultivation (November to February) output is doubled by working two shifts per day. The compost is sold mostly to farmers within a radius of 100/150 km around Cairo, who also pay the transport costs. The operating costs and also part of some welfare projects are paid for from the sale of the compost.

However, not all organic waste reprocessing activities are cost-effective. One of the bottle-necks in organic waste processing is the marketing of end products, among others the marketing of compost. High transportation costs limit the use of compost to the surrounding areas of the city. Urban agriculture could be an option for the application of large amounts of organic waste. Links could be sought with the many urban women who grow and market vegetables. Urban 'greening', that is, supplying green areas (e.g. parks) for the improvement of living conditions, also offers possibilities for the application of compost. So far, these issues have not yet received the attention they deserve and their potential is hardly utilized, also because of the negative image of urban waste.

There is a conflict between the financial constraints and the ecological advantages of resource recovery of organic waste material. Large-scale composting activities, also in industrialized countries, have shown that environmental benefits are more realistic targets than economic feasibility. The question is whether compost production and organic waste recovery in general should be seen as a way to secure profits or rather as a contribution to social and ecological improvements. Organic waste recovery reduces the overall volume of solid waste that needs to be disposed of in sanitary landfills, thus reducing transportation and disposal costs. These so-called opportunity savings could be used to subsidize recovery initiatives to a certain extent.

The improvement of working conditions and the environment

Although the recovery of organic waste has many beneficial aspects, including ecological ones, the reprocessing methods themselves are not always environmentally sound and may pose health hazards to the workers and to the inhabitants, since small-scale, informal activities are often carried out within crowded residential areas.

In urban areas, livestock rearing presents a number of health risks, not only because human diseases can be spread through the waste, but also because of the unsanitary conditions created within residential areas. For these reasons, and because of the foul odours that are often generated, the practice of animal raising is sometimes forbidden in city centres. These considerations pose a dilemma as to whether livestock rearing should be encouraged or not. However, environmental risks should also be seen in relation to the local circumstances. It is of little use to improve living conditions considerably or thus to forbid certain activities, when citizens struggle for their survival and lack access to basic services, such as adequate water and sanitation supplies.

In the centre of Manila, for example, backyard pig raising was once widespread, but as the population grew and space became scarcer, the practice was prohibited. The activity has now been displaced from the centre to the surrounding neighbourhoods, and the number of backyard pig breeders increases with the distance from the centre of Manila. Another option is to attempt to prevent or reduce the risk of environmental and health problems, for example by immediately recycling animal waste into agricultural production, but this again, depends on local circumstances.

Animal raising as carried out in urban areas such as Manila also generates new waste, since the untreated manure is usually allowed to drain into the environment. In Cairo, however, the Zabbaleen take solid pig manure to the nearby composting plant where the mixture of manure and organic waste is sold as compost. The health of the animals is checked at the neighbouring veterinary clinic of GAMEYA (the Association of Garbage Collectors).

Another example of a possible negative side-effect constitutes the quality of the compost. In 1991, a chemical analysis of the compost produced at the plant in Cairo showed that it fell well below European safety standards;

it contained high levels of zinc and lead, and even dangerous levels of cadmium. It was assumed that the organic waste was being contaminated by mixing with non-organic, sometimes hazardous waste (such as household batteries) during storage and collection. Efforts are now being made to avoid this contamination. In an experimental project, 600 households are separating their organic and non-organic refuse before it is collected by the Zabbaleen. The resulting health and efficiency effects on the participating Zabbaleen community, the men who collect the waste and the women who sort the waste in their backyards, as well as the quality of the compost are being monitored. The refuse segregation into dry (non-organic) and wet (organic) fractions could make the Zabbaleens' job easier and less dirty, while they may fetch higher prices for the cleaner 'raw materials'. Improvements, such as separation at source but also precautionary measures for the workers, like the need for protective clothing and face masks, should be taken into consideration as much as possible.

Towards integrated waste management systems

The re-use of organic material as animal or fish feed, compost or fuel can contribute to the solution of urban problems such as the need for income-generating employment opportunities, food production, the lack of adequate waste disposal sites, energy supplies, and maintaining environmental quality. In managing waste collection and disposal systems, these benefits should be taken into account and the various recycling possibilities should be incorporated on both the implementation and policy level.

Resource recovery and utilization are essential elements in any effort to achieve a sustainable level of waste management. Enhancing the recovery of organic waste can restore various natural cycles, thus preventing the loss of raw materials, energy and nutrients. An example of an integrated system of urban waste recovery is the multiple use made of waste at a dumpsite in Calcutta. Figure 1 indicates the various cycles of nutrients and the optimal use that is made of the various waste resources. Organic waste is used both for animal fodder and for growing vegetables. Sewage water flows into fish ponds· and the effluent is used as irrigation water.

The waste recovery system as it is operated by the Zabbaleen in Cairo, including pig raising on organic waste, co-composting of pig manure and garbage and reprocessing of other waste materials such as paper and plastics, is another example of the integrated re-use of organic waste. Both systems, however, could be improved in terms of product quality and working and living conditions.

Waste treatment near the source of generation and separation at source could be other elements of an integrated approach and an important means of preventing the shift of environmental problems to adjacent urban areas, to urban fringes, to more remote places or to future generations. Composting, at household as well as at neighbourhood level, could be part of such a strategy.

Appropriate technology

Until now, transfer of technology has mainly taken place from the industrialized countries to the economically less developed ones, although most of the time these technologies were not directly applicable. The type and com-

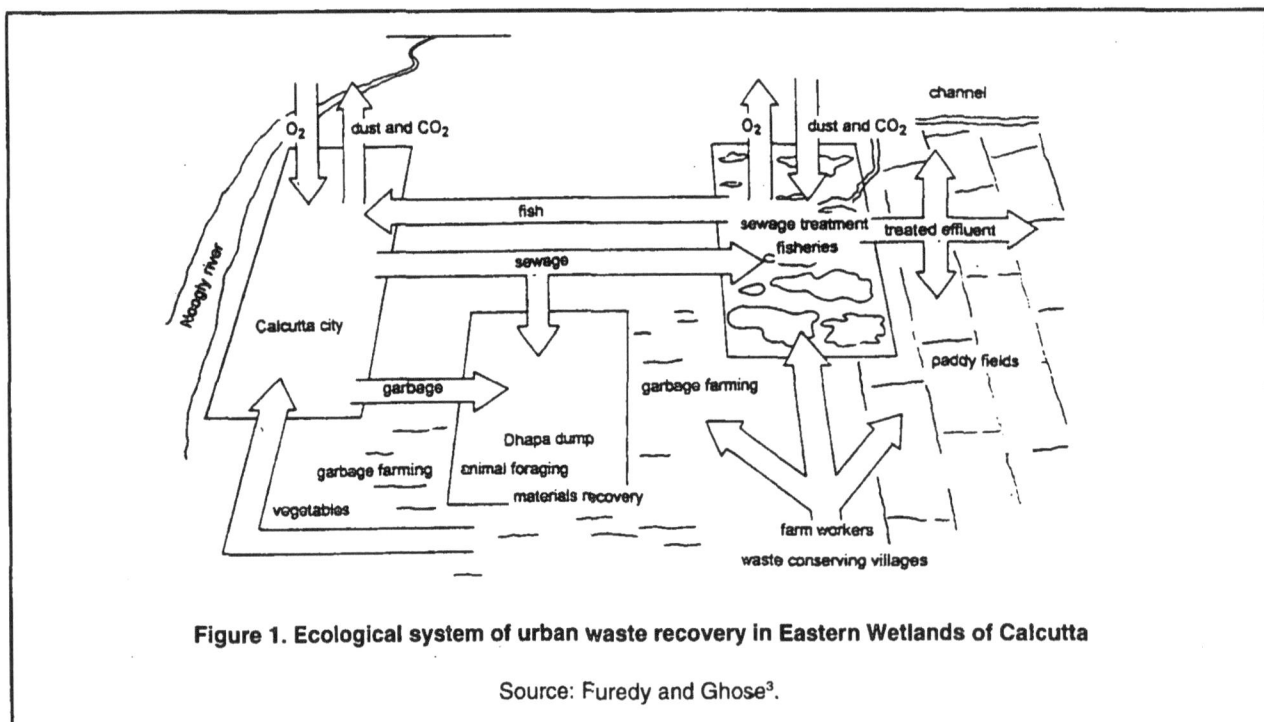

Figure 1. Ecological system of urban waste recovery in Eastern Wetlands of Calcutta

Source: Furedy and Ghose[3].

position of waste, the lack of capital and specific technical know-how, the need for employment generation, the existence of a large informal waste collection sector and cultural attitudes are far from all the aspects that should be considered when developing a sustainable solid waste recovery system in low- and middle-income countries.

Efforts to simply transfer reprocessing techniques based on high-cost equipment should therefore be discouraged. At the same time, one should remain alert to valuable know-how and technologies that could be of use to economically less developed countries. Reprocessing techniques and methodologies, such as forced aeration systems or separation at source that are applied in industrialized countries may provide an option if the aforementioned aspects have been taken into account in the feasibility studies that precede the choice of technology. Transfer of technology consists of more than just technical solutions; a number of financial and social problems should also be solved.

Rather than copying waste management systems that work in the affluent societies of Europe and North-America, countries in Africa and Asia could also look for successful approaches in their own countries. The exchange of ideas, adapted technologies and approaches through a so-called South-South exchange (for example, between Asian and African countries) probably offers more opportunities, and has as yet, not received the attention it deserves.

Many initiatives have been undertaken and many experiments have been carried out in the field of organic waste recovery.

Some activities can indeed serve as examples for other non-governmental initiatives and local government bodies who would like to establish a more appropriate solid waste management system while improving the livelihood of a large number of entrepreneurs involved in waste recovery activities, and are indeed worth being adapted to local circumstances and conditions.

Private initiatives and the role of public authorities

The viability of resource recovery systems depends upon a number of important technical, socio-economic and political relationships. Macro-economic influences such as prices on the international market and trading policies, government policies such as import regulations, and municipal policies also affect the level of resource recovery that will be feasible. For example, the re-use of organic material as an organic fertilizer can reduce the country's dependence on imported fertilizers. However, if chemical fertilizers are cheaper because of government subsidies, farmers are likely to be less interested in using locally produced organic waste as a soil conditioner.

In the past, many municipalities opted for capital-intensive solutions to the waste problem, such as buying large-scale mechanical composting plants rather than developing small-scale, low-tech and low-cost approaches. This is not surprising, since the installation of a plant is an easy decision for a city council, especially if it is offered as a gift or on low-interest financing under a bilateral aid agreement. Also, a people-centred approach requires much more elaborate and decentralized decision-making and coordination. Donor agencies may also play an important role in this process in that they often push certain technologies developed in their own countries, for example compactor trucks or mechanical composting systems.

For various reasons, informal resource recovery, either by micro-entrepreneurs or by communities, has not received the support it deserves. In low-income countries in particular, where unregulated dumping is usually the cheapest means of waste disposing, activities in this field are poorly stimulated and supported by local and national governments. Sometimes municipal policies (deliberately or accidentally) undermine small-scale recovery activities. Different methods of waste treatment and disposal have to be compared, not only in terms of their ecological benefits and economic output, but also of their impact on the less privileged.

A prerequisite for well-functioning organic waste recovery activities is the co-operation between government and private initiator. The wave of democratization processes all over the world facilitates the co-operation between public authorities and private initiatives. This is clearly shown by the Bamako case. Micro-entrepreneurs not only make and sell compost at the dump sites, but several private enterprises and non-governmental organizations have also started waste collection and treatment services. However, to maintain public health and environmental standards the final responsibility (e.g. legislation, co-ordination and control of private services) for waste collection, treatment and disposal should rest with the government.

Some issues that governments could address, depending on their resources and responsibilities, include:

- facilitating the composting of organic waste and other resource recovery processes;
- stimulating the development and implementation of appropriate technologies for organic waste treatment;
- recognizing and integrating the existing informal recycling networks within municipal solid waste management systems;
- formulating policies to protect and encourage the horizontal growth of small-scale resource recovery initiatives;
- creating legal frameworks and controlling mechanisms that will enhance safety in the working place as well as protect the environment;
- stimulating urban agriculture and the 'greening' of cities;
- encouraging the separation of waste fractions at source;
- developing educational material for public information and awareness raising campaigns.

The problem in introducing small-scale resource recovery modules that can contribute to sustainable waste management systems is more a matter of perception than of technology. It requires interdisciplinary co-operation at several levels among various actors, such as municipal and national governments, non-governmental initiators (varying from welfare to women and environmental organizations), research institutes, scholars, community representatives and so on.

Many questions are still to be answered, such as how small-scale resource recovery activities can be optimized under local circumstances and best fit in a broader perspective on waste management. But from the practical experiences gained all over the world, important lessons can be learnt and decisive steps in the adequate direction can be taken.

Linking the insufficient municipal cleansing services to the informal sector services, especially in the field of organic waste reprocessing, could provide a considerable contribution towards urban solid waste removal. The informal sector definitely makes a contribution to a healthy living environment in cities by reutilizing waste materials, since the reuse of organic waste helps to prevent environmental degradation and pollution. Many people depend for their survival on the jobs provided in this sector. Also, organic waste recovery may save foreign currency and natural resources, particularly in the form of raw materials and energy. So, by integrating small-scale organic waste recovery activities in the municipal waste management system, savings can be realized, more employment can be generated and environmental and health conditions can be improved.

1. These consultants are: EQI/Cairo, AUC/Cairo, Undugu Society/Nairobi, Ptr/Calcutta, CAPS/Manila, ABP/Ghana and GERAD/Bamako.

2. A complete overview of the results can be found in *Organic Waste; options for small-scale resource recovery*, which is the first publication in the Urban Solid Waste Series. The book can be ordered from TOOL, Sarphatistraat 650, 1018 AV Amsterdam, The Netherlands. Forthcoming publications are: *Plastics Waste*, *Rubber Waste* and *Hazardous Waste*.

3. Furedy, C. and D. Ghose. 'Resource conserving tradions and the creative use of urban wastes: the sewage-fed fisheries and garbage farms of Calcutta'. *Conservation and Recycling*, Vol. 7 (2 4), 1984.

Decentralized solid waste management approach

Anselm Rosario, Director, Waste Wise, Bangalore, India

BANGALORE LIKE MANY cities in the developing world has outgrown its infrustructure. The Solid Waste Management' an essential urban service, while consuming 40 to 50 per cent of the municipal budget, does not adequately perform' itself posing a variety of economic and environmental problems. While the citizens remain indifferent' blaming the municipal authorities, the backlog of waste remaining uncollected on the road sides also poses other problems. It attracts large numbers of waste pickers who retrieve recyclable waste. Waste picking tends to strew the waste around the bins and exposes the waste pickers to direct contact with decaying materials. Consequently public health and individual health of waste pickers suffer. The waste pickers emerge from low socio-economic groups and live in conditions of utter poverty and deprivation. They have no institutional or state support.

Waste Wise evolved out of a social concern to recognize the jobs of waste pickers, to render their working conditions less hazardous and to gain public recognition for their role in recycling and protecting the environment. It also emerged out of the ecological need to educate the people about their environment, the damage indiscriminate waste disposal behaviours can cause. Waste Wise aimed at the creation of a number of decentralized approaches in the city, where the citizen groups as well as waste pickers can formally be involved in managing waste. It was believed, decentralization would place onus on people to reduce waste, segregate at source and to facilitate recycling. It was also argued, decentralization instead of competing or replacing the formal system of handling waste operated by the municipal bodies, would complement the process.

Waste Wise has more than a decade of experience working with waste pickers on a number of issues related to their survival. One of the biggest problems Waste Wise has constantly to confront the ever growing number of waste pickers every year. A close relationship was identified between waste picking activities and indiscriminate throw away garbage behaviour. While the recyclables found their own outlets, a large component of municipal garbage namely organic materials remained strewn around or uncollected on road sides. It is this portion of the garbage which poses variety of problems and incurs heavy expenditure for disposal.

In order to address the problem at its roots and to quantify hidden nuances in the formal system and waste picking trade, a research study was formulated during April 1990. The objectives of the research was to study formal and informal systems of waste handling and to explore alternatives in collection, transportation and treatment of waste. Promoting community participation, integration of waste pickers and finding out ways and means to improve the conditions of these people were also the considered one of the objectives of the study.

The research which was completed during December 1990, resulted in phase I of a pilot programme at a predominantly middle income locality in Bangalore. Initially the households in the pilot area were sensitized to understand the hazards and the potential garbage. The sensitization was done in various forms starting from discussing with people at the household level, identifying groups of them who showed interest and bringing them together for meetings and planning. Leaflets explaining the need for segregation, what to segregate and how to store the waste were distributed among the 150 households which was selected for pilot scheme. The waste pickers in the area with whom Waste Wise had contact were brought into scheme in the form of exposure to households, training and sharing of information on the operation. A formal launch was done during April 1991 and door to door collection of segregated waste materials was initiated with the help of waste pickers.

One of the options which emerged out of the research study was to convert the organic portion of the municipal garbage into compost, as close to the source as possible, so as to reduce the transport costs. Utilization of the earth worms to hasten the process of composting and also eliminate the malodour associated with decaying garbage through worm activity was considered appropriate for the pilot scheme. A complex interdisciplinary task was undertaken with experts from University of Agriculture, Bangalore (UAS), the Karnataka State Council for Science and Technology (KSCST), with experts in solid waste management and others related to urban planning.

A small piece of land for treatment of garbage on experimental level was obtained from the Bangalore City Corporation (Municipal Body) in a park near pilot area. Phase II of the project resulted in establishment of Vermi Compost Grove in the park area. The purpose was to evolve vermicompost technique suited to Bangalore and study the costs and benefits associated with the scheme.

Today the pilot project operates with 400 households in Jayanagar area. Approximately 250 to 300 kgs of wet organic waste and 20 - 25 kgs of dry waste are collected from the households every day, six days a week. Insanitary wastes are collected once in a week and burned. The dry waste like paper, plastic, metals and glass are stored for a week by the waste pickers and sold in bulk to the regular outlets. The wet garbage is transported from households to park area by hand driven trolleys and dumped in an open pit measuring 3m x 1.5m with a depth of 0.75 m. It is covered with dry leaves during fall or available materials stored for the purpose in the park area. It is then allowed to decompose for ten days with regular turning on alternate days. After 10 days, partly decomposed organic matter is transferred to another closed pit with same dimension as the open pit. The second pit has a roof like structure made out of mild steel to prevent excess sunlight and water during rainy days. The pits are also covered with wire mesh to prevent stray dogs and cows which tend to chew from the pits and are prevalent on Indian streets. The cover prevents the entry of rodents into the pit.

The earth worms of special species cultured by UAS are introduced into the second pit and garbage is kept moist by adding water. It takes about 30 days for the worm activity to be completed with regular turning of the garbage every three days. At the end of four weeks the compost is ready consisting of worm castings as well as finely broken down particles of organic materials. It is then made into mounds and allowed to remain for about 24 hours. During this time the worms move to the bottom of the pit and the compost is skimmed from top. The compost is then sieved using 3mm mesh so as to remove juvenile worms, cocoons and other undigested compounds found in composting process. These are then stored in one part of the park to be used as the covering material for the fresh garbage collected every day. The sieved vermi-compost is then packed in 1 or 50 kgs plastic bags and sold to nurseries, florist, landscapers, home gardeners and small farmers.

Economical/organizational information

Waste Wise utilizes the services of four waste pickers for door to door collection. Each waste picker handles 100 households and is paid directly by them which works out at Rs.5/- per household, per month. The waste picker earns about Rs. 500/ per month (17 US$) plus earnings from the dry waste sales. Two more waste pickers are employed at the site for the compost production @ Rs. 500/- per month plus 10 per cent commission on total compost produced. Waste Wise also employs a community organizer who works in the neighbourhood and with the boys on the street @ Rs. 3000/- per month (100 US$) and a project supervisor who looks after regularity of the collection and compost production @ Rs. 1000/- per month (33 US$).

The compost production per month amounts to 1.5 tonnes approximately which is sold at Rs. 2000/- per tonne (67 US$). The salary of the community organizer is covered by the funding received from the KSCST and Terres des Hommes, Geneva, who jointly supported capital costs, purchase of equipments, pit construction and production of publicity materials. From the compost production two waste pickers who operate at the site and the supervisor are paid. The rest of the money from compost sale either gets accumulated or spent on operational cost like purchase/repair of equipment, laundry costs of uniforms, recreational and medical care activities aimed at other waste pickers who are in the neighbourhood.

Currently, Waste Wise is performing the following functions: promoting decentralised experiments, identifying and training waste pickers, production/sale of vermi compost, promoting environmental sanitation and expansion programme in eight other localities.

The results of the experiment show following benefits

(i) Among the citizens the awareness on garbage issue and willingness to tackle it effectively are high. Other groups have emerged on their own or with support from NGOs.

(ii) Recognition is given to informal labour of the waste pickers, their earnings are better, they work in hygienic conditions and learn ways to convert organic portion into saleable products.

(iii) Environmental benefits in terms of ploughing back the recycled material into soil and subsequent reduction in transportation and disposal costs.

(iv) Since the scheme uses simple technology, indigenous materials and available labour, it could be replicated by residents' group with minimum support.

(v) The scheme can sustain itself in terms of compost produced to cover the salaries and operational costs. However capital costs to start up is necessary.

The problems

(i) It is a land intensive approach and therefore finding land with water facilities within urban set up are difficult.

(ii) Ensuring regularity in collection is essential for the success of the scheme. It takes a lot of effort to train street boys who are habituated to free life style on the street.

(iii) Ensuring 100 per cent segregation is difficult. Currently only 80 per cent cooperation got from the community in terms of segregation and monthly fee payment.

(iv) Any decentralized approach must to some extent depend upon the municipal bodies for its operation. Eliciting co-operation from the municipal bodies in terms of getting land and other support is a mammoth task.

Community based SWM project preparation

Paneer Selvam, Env. Eng. UNDP/World Bank Water & Sanitation Prog., RWSG-SA, New Delhi, India

MOST INDIAN MUNICIPALITIES, despite spending 30 to 50 per cent of their total municipal services budget on Solid Waste Management (SWM), are unable to provide satisfactory SWM services. Recognizing the need to develop a sustainable SWM model, particularly for small towns of India, and at the request of the Government of India (GOI), the Regional Water and Sanitation Group - South Asia (RWSG-SA)[1] is assisting the Government of Goa (GOG) in project preparation, planning and implementation of a community based solid waste management system for Panaji, the capital of Goa. This paper discusses the project preparation process followed by the RWSG-SA and salient features of the proposed SWM system.

Current SWM practices in Panaji

Panaji the capital of Goa, is a small but well developed town with a population of 42 915, as per 1991 GOI census. Panaji with an area of 7.6 sq. km, is popular for its beautiful beaches and historical churches.

Panaji Municipal Council (PMC) is responsible for collection, transportation and disposal of solid wastes generated within the municipal limits. Households and establishments including hospitals, private nursing homes, restaurants, etc., deposit their wastes in communal waste storage bins, for subsequent collection (manual) and transportation to an undeveloped and unsanitary dumping site at Chimbel, 7 km away from Panaji. A large number of waste pickers make their livelihood by collecting a variety of recyclable wastes from bins and the disposal site. Silt from storm water drains and construction wastes are collected separately by the PMC's engineering division for disposal in low lying areas. The current SWM practices need substantial improvements, particularly in the areas of collection and safe disposal of infectious wastes from hospitals and nursing homes; transportation and disposal systems; organizational reforms and optimal use of resources to maximize the manpower and vehicle productivity.

Project preparation process

The five major elements for developing a sustainable SWM project, details of which are discussed below, were studied by a multi-disciplinary project preparation team sponsored[2] by RWSG-SA.

Quantity and quality of wastes

Reliable data on quantity and quality of wastes are important for the design of optimal collection, transportation and disposal options. Detailed field investigations were carried out to measure the quantity (waste generation rates) and quality of wastes from each category of major waste generators such as: households, restaurants, shops, hospitals, markets and street sweeping. Based on these measured values, the total quantity and the physical and chemical characteristics of combined wastes were computed. This information was supplemented with the quantity of wastes recycled by the non-formal sector - waste pickers and waste dealers.

Salient findings of field investigations:

- Panaji generates daily about 22 tons of wastes (refuse) and about 11.5 tons of construction wastes. Households (40 per cent) and restaurants (27 per cent) are the two major waste generators.

- About 1.8 tons, eight per cent of total wastes, are collected daily by waste pickers for recycling.

- Panaji wastes are suitable for composting —70 per cent organic with a C/N ratio of 17; and NOT suitable for incineration — low net calorific value (1300 kcal/kg) and high moisture (65 per cent).

Community needs and perceptions

A social survey, covering about 10 per cent of households and establishments, was carried out to (a) study the current practices on waste storage and community level disposal; (b) assess the community perception of the existing primary collection system; and (c) evaluate community preferences and willingness to pay for improved primary collection services.

The major issues raised by the community are: inadequate number and faulty design of bins; irregular clearing by PMC workers, and the wet and unhygienic conditions around the bins.

Meetings with representatives of other major waste generators, restaurants, hospitals and nursing homes revealed their preference for a personalized 'door-to-door' system and their willingness to pay for the improved service level.

The social survey findings formed an important basis for the development of a sustainable SWM model. For example, the Panaji survey provided some unexpected

---

[1] RWSG-SA is part of the UNDP-World Bank Water and Sanitation Program.

[2] The Panaji SWM Study was financed by a grant from the Government of Norway.

results; people in Panaji are not interested in 'door-to-door' collection of wastes; people are willing to make a monthly payment of Rs.10/- (US$ 1 = INR 31.80 6/94 rate) per household for a communal primary waste collection system with an improved bin design and daily clearance through a mechanized system.

Collection, transportation and disposal

The findings of the field investigations and the social survey helped in identifying possible options to improve the existing collection, transportation and disposal of wastes. Considerations on the use of indigenous technologies; availability of facilities for operation and maintenance (O&M); resource recovery potential; and capital and O&M costs influenced the selection of an optimal option.

Transportation being the major cost of a SWM system, specific unit cost analysis for different options was carried out to evaluate the impact of the improved system on manpower and vehicle productivity.

Various disposal options such as composting, pelletization, incineration, etc. were evaluated. The potential for resource recovery and revenue generation influenced the decision for composting organic wastes from vegetable markets and restaurants.

Institutional needs

The inherent institutional inadequacies that affect the service delivery were analyzed. Issues like fragmented responsibilities, insensitive legal environment, lack of work norms, employment of untrained staff, etc. were specifically addressed, to identify the institutional improvements needed to sustain the proposed SWM system.

Financial management and cost recovery

For optimal utilization of scarce financial resources and to ensure sufficient funds for O&M and capital replacement, the project included options for improving the SWM accounting procedures and cost recovery from house holders, hospitals and restaurants. Discounted cash flow analysis of capital investments and estimated annual recurring costs for different cost recovery scenarios (50 and 100 per cent recovery) were carried out to facilitate the financial evaluation of different options.

Recommendations to improve the existing swm situation

Brief summary of recommendations, evolved as a result of the above described process, is given below:

Primary collection system

- *Replace the existing bottomless cement concrete bins* with suitable numbers of metallic bins, a minimum of one bin within 50 m of all households. The area surround-

ing the bin should be paved with cement concrete or asphalt.

- *Collect the restaurant wastes from 'door-to-door',* twice a day, on a full cost recovery basis. The six existing closed body vehicles should be used for servicing 200 restaurants per day.

- *Collect the infectious wastes separately* from hospitals and nursing homes for incineration at the Goa Government Medical College Hospital, on a full cost recovery basis.

- *Use the existing garbage compactor* exclusively for collecting the organic wastes from the municipal market.

Street sweeping

- *Divide Panaji town into 97 beats,* and employ a minimum of 78 sweepers to work throughout the year.

- *Provide the sweepers with more efficient tools* such as long handle brooms and handcarts with containers, to improve their productivity.

Transportation of wastes

- *Procure five vehicles, fitted with a simple hydraulic device* for direct transfer of wastes from the communal storage bins and tipping, for daily clearance of all the bins.

- *Use the existing vehicles for transporting construction wastes and silt.* Four workers, instead of the 35 employed at present, should be adequate for collecting and transporting construction wastes.

Waste disposal

- *Install a manual composting plant for processing eight tons of restaurant and market wastes everyday.* With proper marketing, it is not only possible to fully recover the production cost of Rs. 325/- per ton, but also possible to make a modest profit. In the neighbouring states of Gujarat and Maharashtra, good quality compost is sold at Rs. 1,300/- per ton.

- *Develop the abandoned laterite stone quarry pit at Talaulim village, into a sanitary landfill site* for disposing of the remaining wastes, which amounts to about 12 tons per day (excluding 1.8 tons of wastes collected by waste pickers). It is estimated that this landfill site will have a useful life period of 15 years.

- *Improve the existing dump site operation* (until the new site is developed) by spreading the waste and covering it with construction wastes using a hired bulldozer on a regular basis.

Institutional framework

- *Provide the PMC's sanitary department with complete responsibility* for handling all solid wastes except silt, including the management of the vehicle workshop. The engineering department, responsible for construction and maintenance of storm water drains, should continue to handle the silt removed from drains.

- *Gradually shift the 'door-to-door' collection, transportation of wastes from restaurants and nursing homes, and operation and maintenance of the composting plant, to the private sector.* However, PMC should clearly define the roles and responsibilities of the contractors, the minimum service levels, the safety measures for handling infectious wastes, the necessary safeguards against delays and inadequacies in the agreed service level, and the direct cost recovery from beneficiaries in accordance with an approved tariff.

- *Reorganize the SWM department* on the basis of work norms and the manpower suggested for the improved SWM service.

- *Sustain the waste pickers contribution to resource recovery,* by organizing them into a formal group with the help of a local NGO, providing them with tools to sort out wastes, raising their status to that of waste collectors, and providing either free or low-cost medical facilities through the state health department.

Financial viability and project preparation

In the last four years, PMC's SWM expenditure has almost doubled and in 1991-92, it spent about Rs. 7.2 million, 40 per cent of its total income, on SWM. With a capital investment of about Rs. 5.5 million for implementing the above recommendations, the annual recurrent costs could be reduced to about Rs. 4.8 million.

Discounted cash flow analysis of the investment and the estimated savings for two different scenarios, (a) with 100 per cent cost recovery for collection of restaurant and hospital wastes and for operating the compost plant; and (b) 50 per cent cost recovery, indicates that the investment pay-back period varies from three to five years and the debt service ratio from 1.67 to 3.10.

On the basis of the comprehensive analysis and attractive rate of returns on investment, PMC has been able to obtain a loan from the Housing and Urban Development Corporation to finance part of the investment cost. PMC has just started the project implementation process with active community participation, facilitated by a local NGO, and the whole system is expected to be in place by mid 1995.

The views expressed in this paper are entirely those of the author and should not be attributed in any manner to the UNDP-World Bank Water and Sanitation Program, the UNDP, the World Bank, the Government of India or any affiliated organizations.

SECTION 5

WATER RESOURCES

Water economy through drip irrigation

Subhra Chakravarty and Lalita B. Singh, India

THE TOTAL CULTIVATED land in India is of the order of 180 million hectares, about 75 million hectares are irrigated at present, by major, medium and minor irrigation projects by both surface and ground water. The annual rainfall varies from 10 cms to 100 cms in various parts of the country. The average agricultural land holding in the country has declined from 2.28 hectares in 1970-71 to 0.68 hectares in 1985-86, although it varies further among various states. Agriculture contributes to nearly 30 per cent of net domestic product and provides livelihood to about 70 per cent of the labour force. The total food-grain production during 1992 was about 146.2 million tonnes, the per capita production being 203kg based on all India average. The production per hectare is about 1368kg. Steady growth is being achieved though improved agricultural technology, implements and practices. The area under non-food crops is approximately one third of that under food crops.

About 40 per cent of the country consists of non agricultural land, which is generally barren and not under forest cover. However all types of land is prone to drought as well as flood, in addition to soil salinity and degradation into semi-arid dryland in the country.

About 175 million hectares of land are threatened by degradation like saline, alkali soil, water logged areas, ravinous and gullied lands, areas under revages of shifting cultivation desertification etc. About 8000 hectares of good arable land are being lost annually due to ingress of ravine. The total flood prone area is about 260 million hectares. The food grain requirement of the country's estimated population of one billion by 2000 A.D. is about 225 million tonnes. Both national and institutional agencies are engaged in formulating various strategies for achieving this target. Optimization of water use in agricultural activity is a major yardstick in achieving the desired production.

The role of drip irrigation

Drip irrigation is thought of as not so much an irrigation system as a total plant support system meaning that water, fertilizer and other necessities are delivered over quite different situations, namely, a large system suitable for a large grower or corporate farmer, a smaller system which could be used by a reliable operation in a low capital, low agronomic skill environment which may exist at a village level in a lesser developed country.

The basic system consists of polyvinyl chloride hose or tube in which the emitters are installed as per spacing provided during manufacturer. Fertilizers and sanitation agents are kept in specified vessels.

The circulating mains consist of 20-25mm PVC pipe extending from fertilizer vessels and run past every in field valve and then vessel. Also passing through every in field valve in the system in conjunction with fertilizer mains in a high pressure clear water main connected to a pressure system, which does not circulate. Injection of fertilizer can be done as per requirement of crop in the field, at single point as well as multiple points. The advantages of the system can be summarized as:

1. Saving of water
Due to localized application of water to the root of the plant, surface evaporation in reduced, run off is decreased and deep percolation loss is avoided resulting in upto 60 per cent saving in water used in conventional irrigation.

2. Better yield of crop
Increase in root length as well as crop yield has been experienced due to slow and frequent supply of water.

3. Saving in labour and energy
Scientific design of the system using principle of hydraulics would require labour only to start and stop the operation and less energy for pumping less water at lower pressure than open field system.

4. Suitable for poor soil
Both light and heavy soils difficult for ordinary irrigation system can be successfully irrigated by this system.

5. Weed growth minimized
Growth of weeds is reduced due to partial wetting of soil.

6. Convenient for cultural practices
The field is always accessible for spraying, weeding and harvesting

7. Less soil erosion

8. Use of saline water
Due to frequent watering, the soil moisture always remains high thereby salt concentration remains below harmful level.

9. Improve efficiency of fertilizers
Due to reduced loss of nutrient through leaching and run off water and localised application of fertilizers, the efficiency is greatly increased.

Description of the drip system
Depending upon the situation the main water pipe is laid out along or perpendicular to length of farm. Sub-main

runs perpendicular to the main and laterals and laid perpendicular to the submain along the row of plants. The mainline and submain pipes are buried at about 30 cms below the ground. The water supply rate is matched to suit the evaporation rates and the type of motor and pump is designed to deliver required quantity of waste to the system at a pressure of about 1-1.5kg/cm². Addition of a filter system either of gravel type or volume type along with the stainless steel wiremesh is necessary to avoid clogging of the emitters. The main waterline is usually chosen of PVC depending on the quantity of water to be handled. The submains are generally of HDPE tubes of 16mm diameter, on which the dripper or emitters are fixed with the help of black LLDPE thin tubings of 4mm internal diameter. One end of the micro tube is attached to lateral and other to dripper. The dripper or emitters are mainly of two types, threaded and pressure compensated. The threaded type dripper helps to maintain uniform pressure and ensures equal supply of water to every plant. The other type ensures uniform supply of water to all plants even in slopes with high gradients even if there is variation in water line pressure. The other components for assembling the system are gate valve, hose collar, start connectors reducers and plugs for main, submain, laterals etc. The system is varied in design such as biwall, plastic emitter, non plastic emitter, microtube etc. without much variation in water use efficiency.

The drip irrigation system ranges are commonly classified as

(a) Surface drip irrigation system
   (i)   Microtube system
   (ii) Pressure compensating drip system
   (iii) Non-pressure compensating drip system
   (iv) In line drip system

(b) Sub surface drip irrigation system
   (i)   Biwall system
   (ii) Turbotape system
   (iii) Typhoon system

The selection of emitter is based on the following characteristics

(a) should be compact, serviceable and inexpensive to keep the system cost low;
(b) should have a relatively low discharge to keep the system cost low;
(c) should not vary significantly with pressure and this will give good uniformity of distribution;
(d) should have a relatively large cross sectional area to avoid clogging problem.

These will help in achieving high system efficiency by means of high emission uniformity, easy flow management and keeping initial and annual cost to a minimum.

Periodic preventative maintenance is the key for the successful working of microirrigations system. The general maintenance includes regular cleaning of filter, check-ing of emitter functioning, wetting pattern and zone, leakage of dripper in proper position and so on.

The design of the system is based on engineering survey of terrain, assessment of water resources, agronomical details, climatological data for computation of evapotranspiration requirements and analysis of soil and water sample. The system design provides mainline design, submain design, lateral design, dripper details, filtration requirement and maintenance schedule etc. Thus each system has to be location specific for deriving maximum benefit, while the components can be selected from the range available in the market.

Experience of drip irrigation in India

An indication of area covered under drip irrigation is presented in Table 1.

The system was introduced in the 1970s and was standardized as well as popularized among farmers during the 1980s, while the coverage grew from 1500 ha in 1985 to above 25 000 ha to the present days. An analysis of drip development in the States of Maharashtra and Tamil Nadu indicate that irrigation water scarcity as well as subsidy provision have contributed significantly to adoption of drip irrigation. Scarcity of labour has also played a significant role in its popularization in Maharashtra. Propagation of this requires evenly paced development in all spheres namely research, extension, raw material availability and processing, fabrication and services sector etc. Keeping this in view, a National Committee on the use of plastics in agriculture was established in 1981, as a central co-ordinating body for propagation of relevant plasticulture applications in the country, which identified drip irrigation as one of the major thrust areas. The Committee has projected 2 mil ha under drip system by the year 2000, subject to a concerted effort by the implementing agencies including the Government of India, state government and financial institutions. In the eighth five year plan period the institutional financing has been estimated to Rs.51 190 mills.

Table 1		
Sr. No.	State	Area under drip irrigation (in ha)
1.	Maharashtra	15 000
2.	Tamil Nadu	3000
3.	Karnataka	2500
4.	Andhra Pradesh	2000
5.	Gujarat	500
6.	Kerala	500
7.	Other States	1000
	Total	24 500

Cost benefit of the system

The cost of drip supply has been estimated to be at the rate of Rs.20 000 per ha. This can be set off against the benefit in irrigating 18 lakh ha of command area in a major ongoing project as follows:

Possible savings	Rs.million
1. Cost of network of field channels	9 000
2. Cost of land	2 000
3. Saving for filling agricultural fields	1 000
4. Land levelling and drainage	3 000
5. Survey of net work	700
6. Saving in cost of main land	10 300
7. Saving in cost of branch land	7 900
8. Saving in cost of distributaries	1 000
Total	35 800

This is claimed to be same as the cost of supply of drips to cover the area.

The benefit of drip irrigation applied to individual crops has been derived through prolonged field investigation. The finding including saving in water has been presented in Table 2.

However it is advisable to adopt specified system for deriving maximum benefit at a particular condition, which should be recommended by the designer or supplier of the system. Some of the circumstances commanding the suitability of a particular type are indicated below:

A) Surface drip irrigation

1) Pressure compensating system — Recommended under drip (Approx all conditions cost Rs.37 000/ha)

2) Partial pressure system — As above compensating drip turbo drippers (Approx. cost Rs.36 000/ha)
3) Microtube irrigation system — Flat lands (Approx. cost Rs.32 000/ha).

B) Subsurface drip irrigation

1) Biwall irrigation system — More choking was observed in biwall (Approx. cost Rs.35 000/ha).
2) Turbo-tape irrigation system — Recommended in fine textured soil with proper care (Approx. cost Rs.38 000/ha).

In spite of the above findings, the constraints such as high initial cost, non availability of subsidy in all states, high import duty on raw material, lack of awareness, lack of research and development absence of countrywide network for manufacturing and distribution etc. are hampering the growth of adoption of this system. Recently Government of India has announced 50 per cent subsidy in all cases of adoption of drip system, which is likely to make it probable to achieve the projected coverage.

Table 2			
Crop	Yield		
	Drip	Conventional	% inc.
Pomegranate	109 000	75 000 nos.	45
Grapes	3250/ha	264 Q/ha	23
Sugar cane	1000/ha	720 Q/ha	38
Banana	825 Q/ha	400 Q/ha	100
Tomato	480 Q/ha	320 Q/ha	50

Low technology drilling

Bob Elson, WEDC, UK

IN MANY PARTS of the world groundwater is the only source of clean water available to rural populations. Hand-dug wells are the traditional means of accessing water but they require considerable investment in terms of time and effort to construct. Allied to this is the uncertainty in locating groundwater which means that a dry well is very frustrating for those involved. Even with successful wells care is needed to preserve sanitary integrity.

An alternative to well sinking is to drill boreholes and these are quicker to produce as well as being easier to protect from pollution. However the downside is the considerable financial investment to purchase and operate a technologically complex drilling rig. Boreholes therefore are usually associated with major assistance programmes and have been beyond the control of most communities. This situation is slowly changing with the availability of low technology drilling methods and equipment that can be used by most people to drill boreholes suitable for handpumps after only a short period of training. A review of these methods and their applicability constitutes the remainder of this paper.

Basics of drilling

No matter what drilling technique is used there are constraints on the process that must be addressed, these including :-

- The strength and degree of consolidation of the rock being drilled governs the energy required to make the hole (weak unconsolidated rocks are much easier to drill than hard consolidated rocks).
- As the hole is deepened debris or cuttings need to be removed (otherwise the tools will become fast in the hole).
- Unconsolidated rocks may need a means of preventing them from collapsing as the borehole is deepened (using either drilling mud or casing).

- For stronger rocks the cutting tools will require cooling and lubrication.

On the basis of these points it is clear that each drilling method must fulfil several functions and that some methods will only work for specific types of aquifer. Because of the use of handpumps in most cases there is only the need for shallow boreholes and it is only the weaker uppermost rock layers that are being penetrated; so energy input requirements are modest.

Percussion drilling

Perhaps the oldest method of borehole drilling is by the repetitive lifting and dropping of a heavy cutting tool which will chip or break rock and thus make a hole. Tools can be of heavy wood with metal cutting edges, iron bars or girders or specially made steel tools all of which are suspended on a rope or cable to enable them to be lifted and dropped quickly. The lifting can be by hand via levers or pulleys or a small internal combustion engine can power a winch or cathead (Figure 1). (McJunkin 1967 [i])

This method works reasonably well in moderate strength rocks but weak rocks or clays absorb and dissipate the impact energy. In hard rocks progress can be very slow. In practice the cutting tools only break the rock and a means of removing debris is required. A bailer (a metal tube open at the top and with one way valve at base) is usually used but in dry boreholes water needs to be added to produce a cuttings slurry that will easily flow past the valve. Depths are only limited by the size of equipment and time available.

Augers

An auger is a tool that is rotated and pushed into the ground to create a hole. The material cut is retained by the tool which has to be removed from the hole periodically for emptying before further progress. It works well in

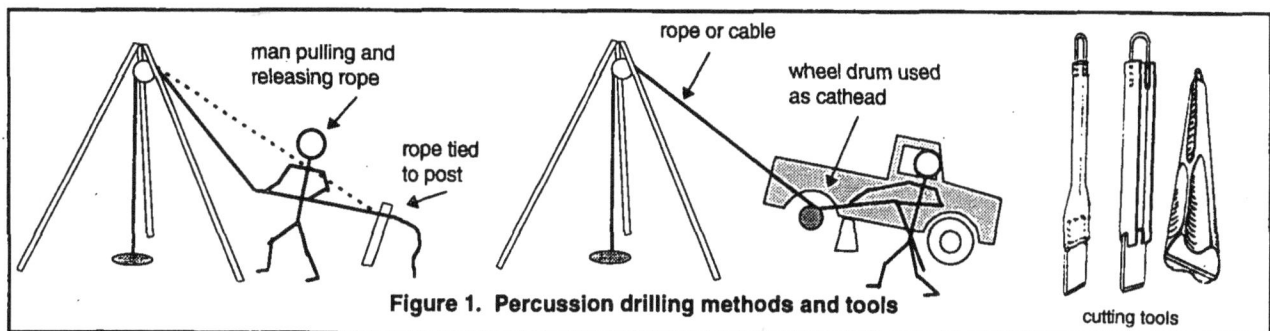

Figure 1. Percussion drilling methods and tools

man pulling and releasing rope

rope tied to post

rope or cable

wheel drum used as cathead

cutting tools

weak unconsolidated to slightly consolidated rocks coping with clays through sands to fine gravel. The main types of cutting head are the bucket auger, the flight auger (archimedean screw thread) Fig 2. but special tools have been devised to cut clays, gravels and to remove small stones. (Brush 1979, DHV 1979, Naugle 1991)

Rotation is by hand and down to about five metres depth the tools and connecting rods can be removed by hand. Beyond this depth a tripod and winch are used to lower and recover the cutting head and drill rods. Fig 3 (Von Elling 1988)

The method works well in soft rocks but below about 15 to 20 metres depth the rock consolidation is such that progress is slow and the rotation effort needed is excessive. Wear and tear on the equipment also becomes high beyond these depths. (Mutwalib 1994) Large rocks or stones at any depth can prematurely stop progress. If the sides of the borehole start collapsing casing tubes must be inserted.

Jetting

In the vicinity of rivers or deltas thick layers of sand and silt can be found. The water table is often only a few metres below the surface and conditions suitable for jetting are found. This method requires water to be pumped down a hollow tube and where it emerges at the base the surrounding sediments are fluidized and then flushed to the surface allowing the tube to descend into the cavity created. Fig 4 The tubes or drill pipe can be made of metal or hollowed out bamboo and the water pump can be hand or foot propelled. In order to assist with the loosening and fluidizing of the sediment a metal cutting edge or bit can be used. An up and down and rotary motion of the drill pipe will help in making the borehole. (US AID 1982, McJunkin 1967[ii]) If the formations are liable to be unstable then the hydrostatic pressure exerted by the water will tend to prevent collapse of the borehole sides.

Drilling depths achieved depend on the thickness and nature of the unconsolidated sediment but 20 metres is common and 80 metres has been achieved. A tripod lifting mechanism is needed for deeper boreholes.

Sludging

This is a variation on jetting and will work in similar circumstances where unconsolidated alluvial sediments are found. Only an up and down motion of a metal or bamboo drill pipe is required usually powered by a hand controlled beam. The drill pipe needs a one way valve often in the form of a hand over the top of the drill pipe (Figure 5).

Starting the borehole from the bottom of a small water filled pit will ensure that the hole is always full of fluid. As the drill pipe is moved it loosens, agitates and fluidizes sediment which enlarges the borehole. On every up stroke of the drill pipe the one way valve is closed (hand placed over top) and water and sediment are lifted from

the hole. Fig 5 This simple action produces a borehole so that casing and screen can be installed prior to installation of a handpump. The method has been pioneered in places such as Bangladesh where depths approaching 80 metres have occasionally been achieved. (Gibson 1969, McJunkin 1967 [ii])

Rotary

Normally rotary drilling methods are associated with large technologically complex drilling rigs but in the past few years small, simple, portable machines have been developed that can drill boreholes suitable for handpump installation in a wide range of rock types.

The cutting action is by the rotation of a drill bit and this is transferred via the drill rods from a small economical internal combustion engine. A hand winch can be used to lift and lower the rods and bit in the borehole. Fig 6 The means of power transfer is a simple robust mechanical linkage that can easily be maintained and repaired without specialised equipment.

For rotary drilling to work the borehole needs to be flushed with air, foam, water or mud. The flush has several functions including removing rock debris, cooling, lubricating and stabilizing weak horizons. Each type of flush has its own particular application but water or drilling mud are the common choice and these are pumped down the hollow drill rods emerging at the drill bit and returning up the annulus between drill rod and borehole. The flush needs to have sufficient velocity to carry cuttings to the surface and this can be provided by a small centrifugal pump driven by an internal combustion engine. Ideally the same model of engine as that on the drill rig. A reservoir for the flushing fluid is needed, this being a pit dug in the ground or tank incorporated into the base of the drilling rig. (Ball 1993)

The important features of these small rotary rigs is that they should be robust, simple to operate and maintain, easy for non-specialists to use, portable and of relatively low cost. A depth capability of between 30 and 40 metres, in most rock types, is also necessary.

Rotary percussion

In situations where strong rocks such as sandstones, limestones, granites and quartzites are present none of the previous techniques will work or be cost effective. However, the use of a down the hole hammer (DTH) will enable boreholes to be drilled quite quickly. This tool is compressed air driven and impacts on the bottom of the borehole very rapidly, whilst being rotated slowly at between 10 and 30 rpm. — hence the name rotary-percussion. Rock is not cut but crushed to very fine debris.

The disadvantage is the need for air compressor but small DTH tools can work with air volumes of 5-7 m3/min (175-250 cfm) and at pressures in the order of 6 bar (100 psi) with the air also being used as a flush. Some of the new portable rotary rigs can easily use a DTH but a simpler alternative is to suspend the drill pipe and DTH

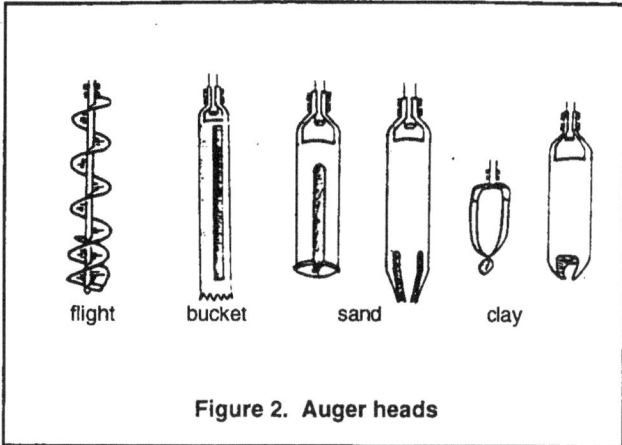

Figure 2. Auger heads

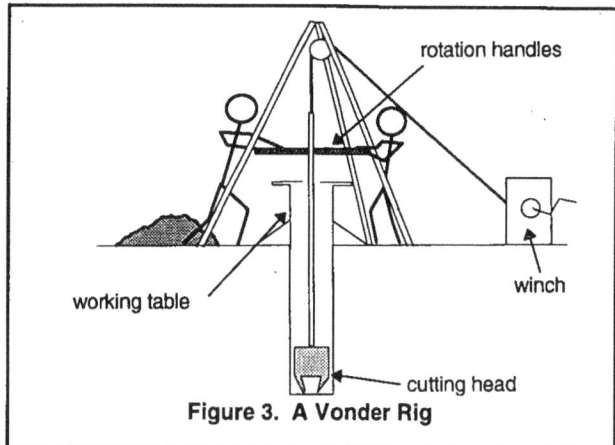

Figure 3. A Vonder Rig

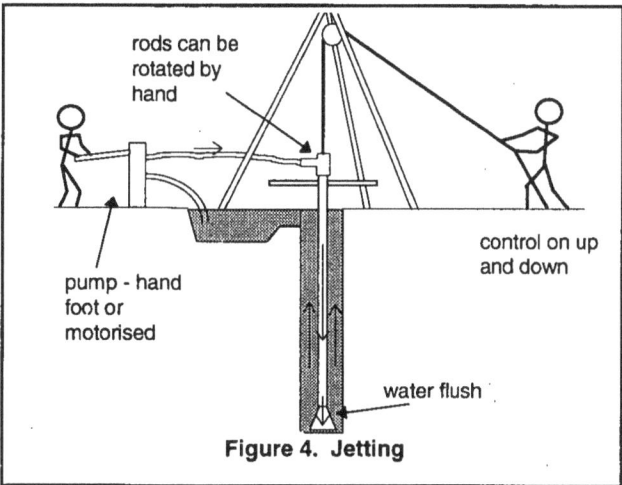

Figure 4. Jetting

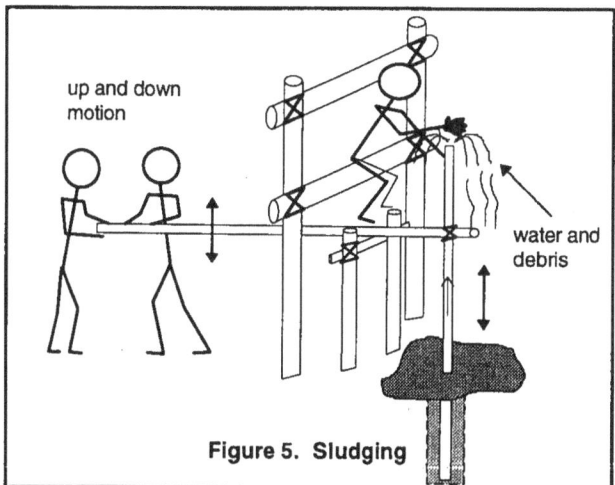

Figure 5. Sludging

Table 1. Advantages and disadvantages

	Advantages	Disadvantages
Percussion	simple, easy to adapt and maintain weak and strong rocks works above and below water table considerable depths	slow, especially in soft clay or unconsolidated deposits and very hard rocks uses heavy equipment and power source needed for large depths problems with unstable formations need to add water in dry holes
Auger	simple, easy to adapt and maintain works above the water table	works only in poorly consolidated rocks hard rocks or boulders stops progress less than 20m, limited penetration below WT problems with unstable formations
Jetting	simple equipment works above and below water table can cope with unstable formations considerable depths	needs a pump and source of water works only in poorly consolidated rocks hard rocks or boulders stops progress
Sludger	simple equipment - locally made works above and below water table can cope with unstable formations considerable depths	needs a source of water works only in poorly consolidated rocks hard rocks or boulders stops progress
Rotary	will drill a wide range of rocks works above and below water table can cope with unstable formations considerable depths rapid progress	complex equipment skill required proper maintenance needed higher costs
Rotary Percussion	will drill the hardest rocks fast works above and below water table considerable depths	complex equipment - air compressor needed skill required proper maintenance required higher costs slow in clays and unconsolidated rocks

from a tripod, use a hand winch for lifting and lowering and provide the rotation by three rig operatives who turn the drilling tools by means of a 'T' bar. Fig 7 This has been quite effective on trials in Uganda. (Allen 1993)

Summary

There are now several low cost, low technology methods for drilling handpump boreholes but each has a particular geological environment where it works best. However the small rotary and DTH rigs are much more adaptable and their relative low cost makes them an attractive purchase for an NGO or a co-operative of communities.

In some circumstances a borehole is not adequate or traditions require a large diameter well. The risk associated with the siting and construction of wells can be considerable and one way of reducing this is to utilise several quickly drilled small diameter boreholes to prove the location of the water and the highest yields.

Progress with this affordable and appropriate technology is one way in which communities can have control over finding their own water resources.

Contacts

Consallen Group Sales Ltd. *Rotary Percussion*
23 Oakwood Hill Industrial Estate, Loughton, Essex, IG10 3TZ, UK. Phone/Fax +44 81 508 5006

Eureka UK Ltd. *Rotary and Rotary Percussion*
11 The Quadrant, Hassocks, West Sussex, BN6 8BP, UK. Phone +44 273 844333. Fax +44 273 846332

Van Reekum Materials bv. *Auger*
115 Kanaal Noord, PO Box 98, AB Apeldoorn, The Netherlands. Phone +55 213283 Fax +55 217937

V&W Engineering Ltd. *Vonder Rig - Auger*
PO Box 131, Harare, Zimbabwe
Phone 64365/63417 Fax 64365

Figure 7. Rotary percussion – hand rotated

References

Allen D.V., 1993, *Low Cost Hand Drilling*, Consallen Group Sales Ltd., Essex , UK

Ball P., 1993, *Technical Brochure*, Eureka UK Ltd, Sussex UK

Brush R.E., 1979, *Wells Construction - Hand-dug and Hand Drilled*, Peace Corps, Washington, USA

DHV Consulting Engineers, 1979, *Shallow Wells*, DHV, Amersfoort, The Netherlands

Gibson U.P., Singer R.D., 1969, *Small Wells Manual*, AID, Washington, USA

McJunkin F.E. (Ed), 1967[i], *Item No 17 - An Inexpensive Truck Mounted Jetting/Driven Well Drilling Rig*, Water Supply and Sanitation in Developing Countries, AID-UNC/IPSED, Washington, USA

Ibid (Ed), 1967[ii], *Item No 15 - Jetting Small Tubewells by Hand*, Water Supply and Sanitation in Developing Countries, AID-UNC/IPSED, Washington, USA

Mutwalib W., 1994, *Evaluation of the Muyembe Rural Water Supply*, Diploma Project, Loughborough University of Technology, UK

Naugle J., 1991, *Hand Augered Garden Wells*, Lutheran World Relief, Niamey, Niger

USAID, 1982, *Technical Note No RWS 2P2, Selecting a Method of Well Construction*, Water for the World Technical Notes, USAID, Washington, USA

Von Elling E.H.W., 1988, *Instructions for Drilling Tubewells with the Vonder Rig*, V&W Engineering Ltd., Harare, Zimbabwe

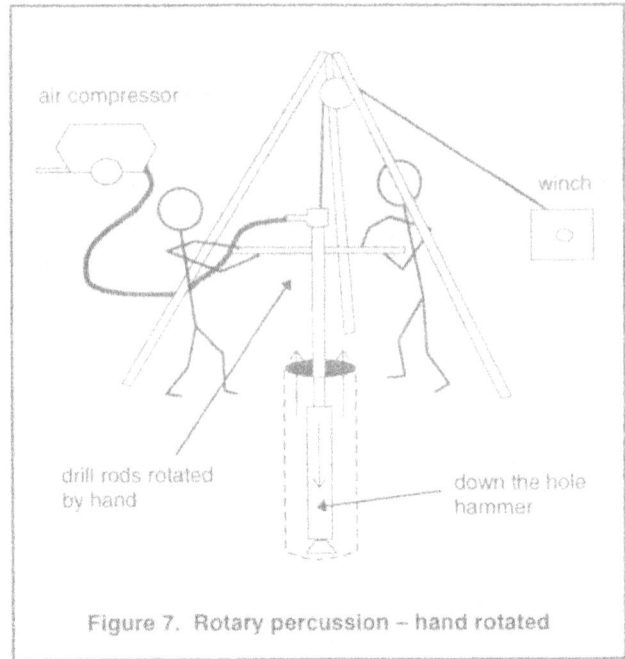

Figure 6. Eureka port-a-rig

Groundwater depletion due to agrowells

K.L.L. Premanath and Tissa Liyanapatabendi, Sri Lanka

NATIONAL WATER SUPPLY and Drainage Board with financial and consultancy (COWIconsult) assistance from DANIDA carried out a study to prepare a Master Plan Report on water and sanitation needs of the Anuradhapura district during February 1992 to March 1993. Anuradhapura district is located in the dry zone of Sri Lanka having an average annual rainfall of about 1200mm. Main occupation of the people in the district is agriculture.

The ground water in Anuradhapura district is presently being used for both domestic and irrigation. Water for irrigation is pumped from agrowells. However, the recharge of groundwater from infiltrated rainwater sets the limits for how much water that can be extracted annually. There were 5720 agrowells constructed within the district by the end of year 1992. Out of these an estimated 2500 were private wells whereas the remaining wells have been constructed with government subsidies. A detailed study was carried out to monitor agrowells in the district with respect to the fluctuation of water levels and varieties in the chemical qualities.

Methodology used in the study of agrowells are (a) Field survey of agrowells collecting technical and agricultural data (b) Interviews with agricultural authorities and district administration (c) Establishment of a data base (d) Analysis of data.

Some 179 agrowells, covering the whole district were identified for monitoring purpose. Data collected from wells show that a typical agrowell is 6.4m deep, has a diameter of 5.4m and a mean water level 3.5m below ground level. Over the year the water level varies between 1.9m and 5.0m below ground level. It is constructed by the owner by hand/mechanical digging and is lined on the inside by cement blocks or bricks. Construction cost excluding own labour and pump has been about Rs. 29 000.00. Pump capacities vary typically between $27m^3/hr$ and $45m^3/hr$.

The Government of Sri Lanka commissioned the agricultural development authority (ADA) through the Ministry of Agriculture Development and Research to implement a nationwide agrowell programme. Hence, the ADA started implementation of an agrowell programme in Anuradhapura District in 1989. Although the aim of the programme has been to assist farmers to cultivate crops at and around their homesteads and in highland areas where cultivation depends on rainwater, it is however now seen that farmers establish agrowells to support paddy cultivation in areas, where it is felt that gravity irrigation from the tanks needs supplementing.

In Anuradhapura district three different implementation modes are in existence. They are implementation of the nationwide programme supported by either ADA, the programme supported by the provincial council of the North Central Province, and there is a substantial number of agrowells being established privately without relation to the official programmes. The number of agrowells and the plans for further well construction at the end of the year 1992 are given in table 1.

It appears from above that a substantial number of agrowells have been already established. Further, all trends point in the direction of a considerable economic success for the programme and it must be assumed that wells may be established at a minimum, at the same rate as planned for 1993.

The beneficiaries of the programme are to a large extent the poor peasants who used to cultivate a single season (Maha) with rainwater and abandon the Yala cultivation due to inadequacy of irrigation water (Ariyabandu 1989). Even the maha crop was subjected to severe water shortage towards the end of the season. Today these farmers are benefitting from full cultivation seasons and in a few cases even a short third season between two major seasons. The cultivation comprises cash crops such as chillies, Bombay onions, red onions and vegetables as indicated by the field survey. Farmers with agrowells have increased their income substantially and in some cases up to ten times. Peasants who often had to migrate as labourers are now full time employed in their cultivation areas. Incomes in the range of Rs 20 000.00 to Rs 40 000.00 per season per acre have reportedly been generated.

The development of agrowells takes place in a rather haphazard way without a general assessment of the hydrogeological environment, the possible yield and a rational siting.

In order to assess the sustainability of agrowells and the possible conflicts with other water demands a rough evaluation of the irrigation water requirements was made based on the most common cultivation such as chillies and Bombay onions. Water requirement from agrowell calculated on the basis of crop evapo-transpiration figures from development of agriculture and crop factors at various growth stages is given in table 2.

Sustainability of the system of lift irrigation using agrowells as the source depends on the recharging of the groundwater resource and the variations in recharge over the years. The recharge depends fully on the hydrological regime and the soil conditions. The recharge may differ

Table 1. Present number of agrowells constructed and planned (Ada, Anuradhapura district)

Year of construction	Agricultural development authority	Provincial council	Private (estimate)
Before 1990	100	159	
1990	159	238	
1991	919	340	
1992	1005	300	
Plans for 1993	1000	400	
Subtotal	3183	1437	2500
TOTAL		7120	

from the highland areas to the valley bottoms, where the paddy cultivations are found.

Potential conflicts are many when there is an over-exploitation of the groundwater resource either on the local or on the regional scale taking place. Indiscriminate groundwater withdrawal will cause depletion of groundwater reservoirs. This will first be observed as a falling groundwater table, a perceived need to deepen the agrowells and a high frequency of wells running dry. The agrowells will interact and the deepest will deprive the other wells of water. Drinking water wells will likewise run dry and natural rivers and streams will suffer from extended dry spells.

Master Plan Study calculations show that for each acre irrigated there has to be 34 acres where recharge takes place in order to have a sustainable situation in highland areas. Similarly for low land areas corresponding area is 17 acres. The table 3 summarises the calculated recharge and the consequences in terms of maximum numbers of standard agrowells.

Under the assumption made, it appears that safe number of agrowells has been vastly exceeded as most development has taken place in the highland areas. It must be expected that serious depletion problems will occur and that agrowells will frequently run dry. They will affect each other in such a way that the deepest wells will make the other wells run dry. Further, nearly domestic water wells will also be drying up. There is thus an urgent need to bring the situation under control in order to avoid a situation like that in Tamil Nadu, where indiscriminate sinking of agrowells has caused a permanent depletion of the ground water resource. From the results of the data collected for the Master Plan, the figure 1 shows the decline in ground water level if present level of construction and use continues and if new construction of agrowells is stopped (Freeze).

Based on the very serious consequences that the continued pumping from the agrowells will have, the situation must be brought immediately under a tight control. Presently there is nothing to stop a farmer on private land from exploiting the national groundwater resource at his will. The following immediate steps are recommended.

o   An urgent freeze of subsidies.

o   An urgent programme of registration of the details of all agrowells, location, the irrigated area, pumping, etc.

o   An urgent implementation of a ground water level monitoring programme in areas with clusters of agrowells.

o   Immediate enforcement of criteria for distance between wells.

o   Identification of legal framework for intervention in particular regarding privately constrcuted agrowells.

o   Implementation of a regulatory system of water rights.

o   Attaching responsibility for allocation of water rights to an independent agency which should also act as a supervising agency.

o   Draw up standard construction details for agrowells that limits the rate of groundwater abstraction.

o   Agricultural agencies should be instrumental in advising farmers on the optimal use of the water withdrawn and should collect information on the construction and use of agrowells.

Reference

Master plan for water supply and sanitation Anuradhapura district, June 1993.

Table 2. Water requirement from source (AGROWELL) for most common crop (figure in mm)

	MAHA	YALA
Crop water requirement A=ETxkc	703	916
B =Application efficiency	55 %	55 %
C = Conveyance efficiency	80 %	80 %
Water requirement at source D=A/B/C	1597	2082
R = Rainfall	970	477
E = Rainfall efficiency	70 %	70 %
Effective rainfall F=RxE	679	334
Water requirement at source adjusted for effective rainfall W=D-F	918	1748

Table 3. Summary of recharge and safe number of wells assuming a cultivable highland area of 30 000 ha (75,000 acre) where agrowells can be established and that agrowells are desirable also within the presently irrigated lowland area.

	Highland (30 000 ha)	Lowland (Irrigated are 110 000 ha)
Recharge	54	109
Required ratio between recharge area and irrigated area	34	17
Maximum no. of standard agrowells (0.81 ha)	1090	7971
Present no. of standard agrowells (0.81 ha)	7,120 (4272 in highland and 2848 in lowland)	

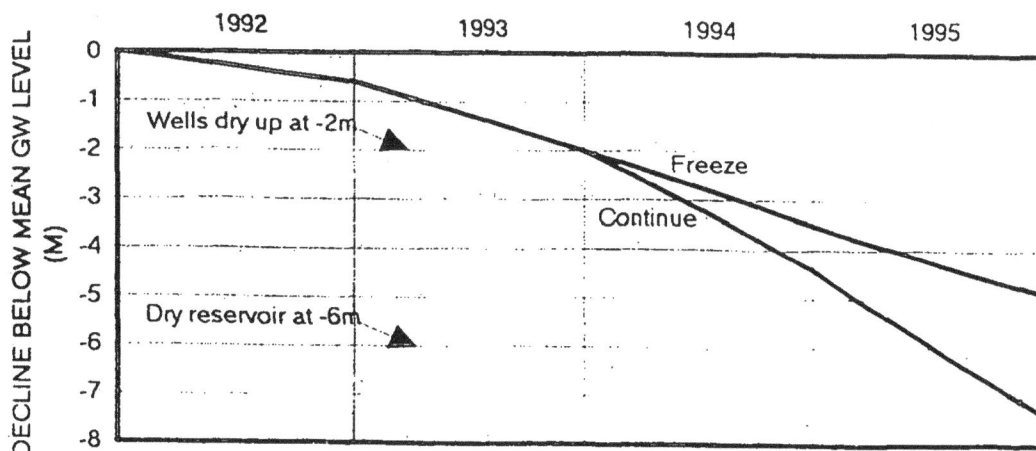

Figure 1. Decline in ground water level of construction and use continue and if new construction of agrowells is stopped (freeze)

Disseminating ram-pump technology

Dr. Terry Thomas, Warwick University, UK

FOR BOTH IRRIGATION and domestic supply, gravity feed is not always possible: water often needs lifting. The power to lift a flow of water can conveniently be expressed as

$$power = \frac{constant \times mean\ flowrate \times height\ lifted}{duty \times efficiency}$$

where 'duty' is a time fraction (pumping hours per day) and 'efficiency' is a product of the efficiencies of the hydraulic circuit, the pump and the prime mover. Pipes are sized to give tolerable hydraulic efficiency and pumps are chosen to match the hydraulic conditions and the energy source available. Duty can also be varied to achieve better matching of the prime mover to the hydraulic circuit: high duties such as continuous 24-hour operation result in low power requirements and cheap piping (see Box on next page).

Whilst in general the power for water-lifting can come from engines, electrical mains, animals, humans or renewable (climatic) sources, in the particular context of rural areas in poor countries the choice is more constrained. In many such countries there are virtually no rural electrical mains, engines pose problems of both fuelling and maintenance, draught animals may be unavailable or difficult to apply to water lifting, renewables are erratic, complex and import intensive. Therefore human-powered lifting and transporting of water is still

common, despite the very high cost of human energy (US$ 2 to 20 per kW hour).

Of the renewables, water power has the longest history, and under favourable conditions is the easiest to use. Several Asian and Latin American countries have developed the capability of building hydro-power systems. Although sites where power can be economically extracted from falling water are rather rare, they generally occur in the same terrain (mountainous) as the greatest water-lifting needs. The use of water power to pump water is therefore an interesting option. Figure 1 shows the main ways of doing this and illustrates the relative simplicity of the hydraulic ram-pump system. A typical such system is shown in Figure 2.

Ram-pumps (invented 200 years ago) are still manufactured in over ten countries and were once commonplace in Europe, The Americas, Africa and some parts of Asia. They have however been largely displaced by motorized pumping in richer countries, whilst in developing countries their use is concentrated in China, Nepal and Colombia. Ram-pump technology is not trivial: designing systems that are reliable, economic and durable (e.g. against flood, theft, silt ....) takes some experience. Generally, in rural areas of developing countries, this skill has been lost since about 1950, and the intermediaries that used to connect ram-pump manufacturers to pump users have disappeared. Old systems lie broken for lack of fairly simple maintenance: new systems are few.

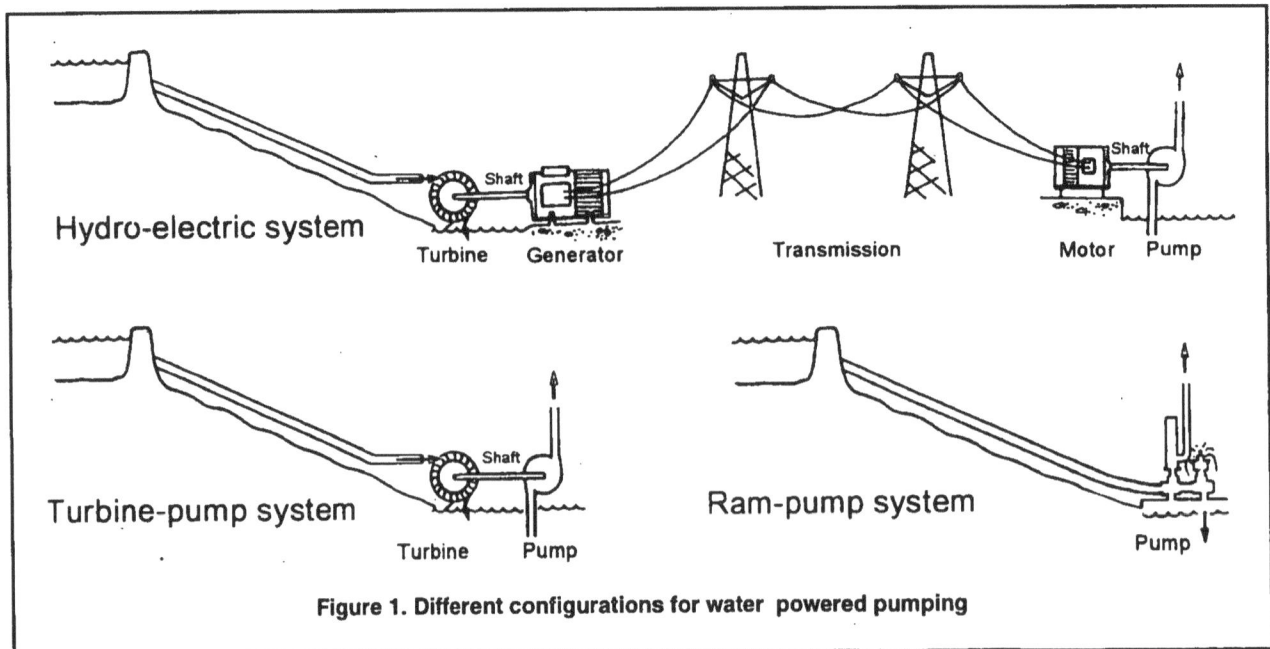

Figure 1. Different configurations for water powered pumping

For various reasons, discussed later, the potential for using ram-pumps seems to be increasing worldwide. Working, primarily in Africa, since 1985 the Development Technology Unit of Warwick University has identified several obstacles to this potential being realized, and has been trying to remove them. This paper records that experience.

The niche of the ram-pump

In suitable terrain, ram-pumps can be used to provide low-power unsupervised pumping. Typical individual ram-pumps can deliver 10 to 200 watts for lifting water; several small pumps can be operated in parallel to feed a single delivery pipe, larger pumps are available from some manufacturers. The power requirements of rural water lifting are illustrated by the following examples, which all assume pipe head losses are 10 per cent of lift. The powers quoted are 'water watts' assuming 24 hours pumping.

domestic supply to a prosperous house (500 litres per day lifted 75m)	5W
village supply (10 000 litres per day lifted 50m)	62W
irrigated garden (0.5 hectare) (35 000 litres per day lifted 20m)	87W

As the ram pump's system efficiency including its drive pipe is 50 per cent to 75 per cent, the hydro-power inputs for the examples above need to be up to twice the figures shown. The ram-pump is therefore well power-matched to these applications. These inputs are obtained at comparatively low drive heads — typically 10 per cent of the delivery head — so the drive flows to ram-pumps are typically twenty times their delivery flows. (In the examples above the drive flow would be typically 7, 140 and 500 litres per minute respectively). This high flow requirement is clearly a constraint on location. On the positive side, however, no ram-pump user can extract more than a small fraction (e.g. five per cent) of any source flow, the bulk of it being passed on downstream to other users: this has some social advantages.

Three other technical constraints require mention. Firstly there is only a limited range of head ratios (delivery height divided by drive head) of 5 to 30 over which a ram-pump is efficient and economic. Secondly neither drive head nor delivery head should exceed the particular pump's rating (often 20m and 100m respectively, but much less for cheap plastic ram-pumps). Thirdly it must be acceptable that the water lifted is derived from - and hence is of the same quality as - the drive water: a ram-pump cannot derive energy from a dirty stream to pump water from a different (cleaner) source.

Disregarding social and organizational factors, we can therefore describe the technical niche of the ram-pump as moist hilly rural areas where there is no mains electricity but a need for lifting water from streams or

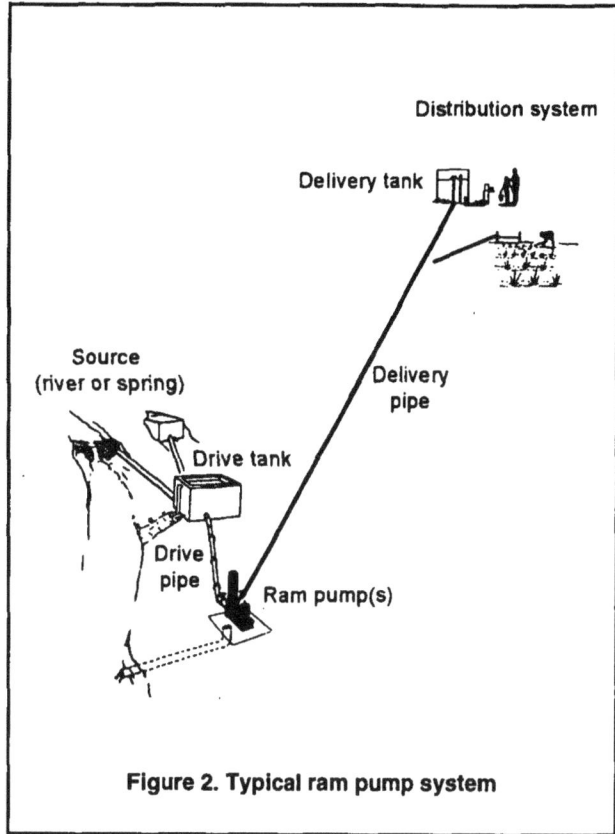

Figure 2. Typical ram pump system

springs. The source must be of adequate quality and have a flow many times that to be lifted.

The problem of minor technologies

One of the more accessible concepts from 20th Century physics has been that of 'critical mass'. If the mass of a radioactive material or the size of an organization is below some threshold its activity dies away; above that threshold the activity sustains itself and may even grow. For most technologies there is similarly a critical scale of application below which the activities needed to sustain it may die away. Such activities include manufacture of components, training of new users and specialist maintenance.

In the case of ram-pump systems, specific skills are needed in manufacture, system design, installation and operation. The skills are not especially high and overlap those needed to manufacture, install etc. other devices. Sometimes such skills are preserved in inanimate form. Thus many ram-pump manufacturers employ steel castings whose foundry patterns were made decades ago. Documents preserve design procedures. Existing installations are available as models for new systems. The critical throughput to sustain commercial manufacture is perhaps 50 pumps per year, it is usually achieved via selling into more than one country. A throughput of only one or two new systems a year might sustain system design and installation skills in a general water contractor. However, a specialist installer might need to put in at least 20 pumps a year to survive.

Continuous pumping versus discontinuous pumping

Water and wind powered pumps, and some electric pumps, are best operated for 24 hours per day. Solar, human and animal-powered pumps are limited to about 8 hours per day. Diesel pumps are typically run for only 2 hours per day in rural areas because they are usually over-powered for their applications. These differences in duty (load factor) have implications for pipe and storage costs. Pipes are sized so as to give a 'tolerable' friction head loss (FHL). What is tolerable depends on the means available to supply this head loss, for example pump power or the slope of the pipe. We know that, for a given length of pipe, FHL is proportional to Q^2/D^5, where Q is flowrate in the pipe and D is its diameter. Also pipe cost (per meter) is typically proportional to D^2. These relationships give the table below, which is based on a specified daily flow.

	Pumping for 24 hours/day (taken as datum)	Pumping for 1 x 8hrs/day	Pumping for 2 x 1 hr/day = 2hrs/day
Power to overcome FHL	1	x27	x1728
Energy to overcome FHL	1	x9	x144
Pipe D for constant FHL	1	x1.5	x2.7
Pipe cost for constant FHL	1	x2.4	x7.3
Typical storage + daily throughput	0.4	0.5	0.4

In reviving an old technology or introducing a new one, the 'critical mass' throughputs need to be estimated. If they are higher than the area of sales or of installer operation can sustain, any intervention to promote the technology will ultimately fail. More important, if the likely demand is thought to be close to such a threshold of sustainability it is worth effort to lower the threshold.

With the technology of hydro-electricity we are used to having separate organisations making turbines, designing systems, building them and operating them. Maintenance may require a fifth agency. Even though some of these organisations operate internationally via local agents, such complexity entails uncertainties that tend to raise the critical size for each of them. Micro-hydropower utilisation has lagged behind its apparent economic potential for these reasons in most countries. Ram-pumping faces similar difficulties.

Often there is a key agency that effectively leads the others involved in a technology. For example a manufacturer of equipment may set up training for its installers, users and maintainers; alternatively a consultant may co-ordinate and supplement the existing skills of the other parties. A low value rural technology does not lend itself to the latter approach.

Experiences in Africa

The author and his DTU colleagues have been trying to revive ram-pump usage in Africa since 1985. An early analysis suggested that foreign (e.g. European) manufacturers selling a few pumps a year via agents could not and would not provide adequate training for local installers. Moreover imported pumps are expensive and difficult to source spares for. In colonial times there were few technical alternatives for water lifting to plantations,

mission hospitals and large schools and it was worth the cost of bringing a ram-pump installer from another continent. Today that is an unacceptably expensive option for a village or farm needing pumped water or for a small-scale pumped irrigation scheme.

In the absence of a design consultant (again unlikely for this scenario), the options for sustainability appeared to be

either to build up the design capability of installation contractors

and/or to encourage local manufacture by an organisation also capable of providing back-up to installers.

The DTU chose the 'and' option, first spending several years in developing simple and cheap pump designs suitable for provincial manufacture and codifying system design and installation procedures. Since 1990 the DTU has been training both producers and installers from nine African and one Asian country, usually using its demonstration centre in the Eastern Highlands of Zimbabwe. There is an ongoing debate about what is the right level of manufacturing technology (hand tool, workshop with electricity, factory), whether manufacture and installation should be undertaken by the same organization, whether low-lift irrigation or high-lift water supply should be given priority, whether installer training should be directed towards governmental, NGO or private organizations and what fraction of possible sites are 'easy' sites suitable for beginners to tackle.

The results have been mixed. Easy sites (with modest lifts, plentiful water, favourable stream geometry and well-organised customers) are perhaps only a few percent of technically feasible sites. The process of system design has proved intimidating to technicians for whom even sizing a pipe for gravity flow is at the limit of their understanding. The input of (expatriate) man and woman power to bring an installation organisation up to the level of competence and confidence to stand alone with this technology has been expensively high. The 'successes' have been with unusually well-resourced NGOs. Commercial manufacture, for example in Kinshasa (Zaire) and Mutare (Zimbabwe) has been started but self-sustaining manufacturer-installer arrangements have not been developed. Of some 30 pumps installed, too many have been 'demonstrations' rather than built to meet real water needs.

Clearly training on courses alone is not enough. Installers and manufacturers need to be visited and helped/ encouraged with production of their first systems. A ram-pump has a certain 'something-for-nothing' magic about it that impresses onlookers and causes any installation to yield many enquiries from neighbouring villages or farms. However the technology's uncertainties, using very cheaply produced pumps in the hands of novice installers, makes it much easier to apply to individual 'rich' farms or institutions than to villages or communal dry-season gardens.

Ram-pump technology has a fascination for enginers and users out of proportion to its current commercial importance. The DTU's 1992 book on system design must have sold more copies worldwide than there have been new systems built! A 1993 day school on ram-pumps in Sri Lanka attracted fifty engineers but so far has resulted in no new systems.

Prospects

Ram-pumping will never be a major technology comparable with motorized pumping from rivers or hand pumping from boreholes. Its particular niche is described above: worldwide there is a potential for between perhaps 10 000 and 200 000 systems. Much of that potential lies in areas where there are currently no system design skills. Availability of pumps need not be a major problem (despite the DTU's local manufacture strategy in Africa), since even though good imported machines cost over $10,000 per kilowatt the pump itself rarely accounts for more than 40 per cent of system costs.

Certain trends worsen the prospects for ram-pumps. Worldwide, water sources are becoming both dirtier and weaker. Some historical ram-pump systems no longer operate because of declining drive flow. Clean spring water is usually associated with very low power levels — in Rwanda for example, the DTU had to design for 80 metre lifts from drive flows under 10 litres per minute, which is on the limits of the technology.

Factors increasing likely demand are the movement of rural populations uphill (under population growth pressures), the expansion in micro-irrigation, the introduction of local ram-pump manufacture (especially in South America) and the availability, apparently for the first time in decades, of both trustable handbooks and training courses.

In Africa the prospects for ram-pump usage seem to depend largely on the confidence of potential installers. Despite much individual innovation there, Africa is not a continent where organizations readily take risks with unknown technology. Elsewhere in the developing world continuation of the current slow expansion of ram-pump usage will depend upon developments in photo-voltaic pumping, its most immediate rival.

The scope for technical improvement of a simple device already used for 200 years is rather small. However, modern materials may permit the pressure vessel (required to smooth the pulsating flow through the delivery valve into a steady flow up the delivery pipe) to be replaced by a pressured bladder. This will allow pumps to be operated slightly under water which has advantages for both efficiency and reliability. Understanding of the causes of erratic pump behavious and of inefficiency is now better than in the past, which designers of pumps and 'trouble-shooters' of systems can draw upon. It is not possible to totally design away temperamental behaviour, during for example system start-up, but its incidence can certainly be reduced.

For the ram-pump to fully occupy its niche, efforts must continue both to simplify the design of reliable systems and to propogate design skills. Although water-powered pumping will never attain the simplicity of drop the suction pipe in the stream and switch on that motorized pumping offers, as users of a renewable energy source, ram-pumps may have time on their side.

References

The following books explain how ram-pumps work and provide system design assistance. They also contain addresses of (Northern) manufacturers.

Meier V. *Hydram Information Package*, SKAT, St. Gallen, 1990.

Knol H. *The Fall and Rise of the Hydraulic Ram-pump*, Drachten (Netherlands), 1991.

Jeffery T. et al *The Hydraulic Ram-pump*, London, IT Pubs, 1992.

Please contact the DTU, Warwick University, Coventry, CV4 7AL, UK for drawings of pumps designed for local fabrication.

SECTION 6

WATER SUPPLY PROJECTS

Leakage control — the neglected solution?

Geoff Bridges, Mott MacDonald, UK

THE MAJOR COMPONENT of unaccounted for water (UFW) in most water supply systems is invariably found to be physical leakage from the distribution system. Not only does this translate into a major loss of revenue for the water utility, but high leakage levels may lead to low system pressures or even intermittent supplies. In some systems water may have to be systematically rationed in an attempt to distribute the available water more equitably. Effective control of leakage and UFW is therefore critical, but frequently it is given insufficient priority.

The range of reported values of UFW in selected countries and major urban systems is very large, as can be seen in Figure 1 (IWSA, 1991; ADB, 1993). Typically, UFW levels in the more developed world are between 15 per cent and 30 per cent, but elsewhere they are more likely to be in the 30 per cent to 60 per cent range. There are exceptions to this, however, notably Singapore with eight per cent UFW, and Dusseldorf in Germany with three per cent (Lackington, 1991).

A feature of all systems that exhibit low values of UFW is that an on-going active leakage control policy is being implemented. This is usually based on waste metering or district metering, supplemented where appropriate by pressure control. In Asia, Tokyo, Hong Kong and Singapore implement waste metering policies whereas in UK most water utilities are now converting to a policy of continuous monitoring. This is effectively a district metering policy integrated with a telemetry system to provide continuous scanning facilities and dynamic simulation of the network. Implications of leakage control on overall system performance are demonstrated in Figure 2..

It is a generally recognised fact that the vast majority of leaks and leakage occur on service pipes to consumers properties. In Tokyo 90 per cent of leaks located, accounting for an estimated 80 per cent of distribution system losses, are attributed to defective service pipes (WSA Seminar, 1994). A similar pattern has been found during recent studies in Hong Kong. Recent UK data indicates that service pipe leakage accounts for up to 45 per cent of total leakage in some UK water distribution systems (OFWAT, 1993.) The high incidence of service pipe leakage has prompted some utilities and countries to supplement active leakage control policies with a comprehensive annual mains replacement or rehabilitation programme, with service connections being replaced at the same time as the distribution main to which they are connected. The banning of inappropriate pipe materials, for example unlined galvanized iron mains, and the

adoption of newer materials such as polyethylene, which have superior hydraulic and corrosion resistance properties, has made a further contribution towards reducing leakage.

In Tokyo, overall annual mains replacement rates of 1.1 per cent have been achieved consistently, with the emphasis placed on cast iron and asbestos cement mains which are being replaced at a rate of two per cent annually. Stainless steel is the preferred material for service connections. In Singapore most unlined cast iron mains have been replaced together with 76 000 service connections. The annual mains replacement rate is 0.5 per cent (Water Malaysia, 1992). The average age of the pipe networks in Japan and Singapore is therefore relatively young. In some German and Dutch systems annual replacement rates of up to three per cent are not uncommon, the average age of such systems being of the order of only 25 years, influenced by major infrastructure reconstruction following World War II. Minimal leakage control effort is expended in systems where high renewal rates are achieved. There is very little published evidence to support the economic justification for such high annual renewal rates which, although leading to low leakage levels, are reflected in considerably higher water charges. The philosophy appears to be that the achievement of low levels of leakage is alone sufficient justification. High replacement rates can only be entertained in relatively affluent parts of the world.

The question therefore remains as to what can be practically done to reduce the high levels of UFW and leakage that characterise many systems in the developing world. Active leakage control policies tailored to the local situation have been proven to be cost-effective throughout the world. The skill lies in selecting the optimum policy and determining the amount of effort to be put into its implementation to provide a positive economic balance, i.e. the cost of policy implementation must be less than the value of the water saved. However, despite the cost-effectiveness of leakage control programmes political considerations may influence technical solutions to improve service levels.

For instance, instead of implementing a cost-effective leakage control programme to improve supplies to consumers, emphasis may be placed on a higher profile capital works scheme to demonstrate that the utility is taking positive action to tackle the problem. However, the construction of a new treatment plant to increase the supply of water to deficient areas is only likely to provide temporary relief unless the underlying leakage problem

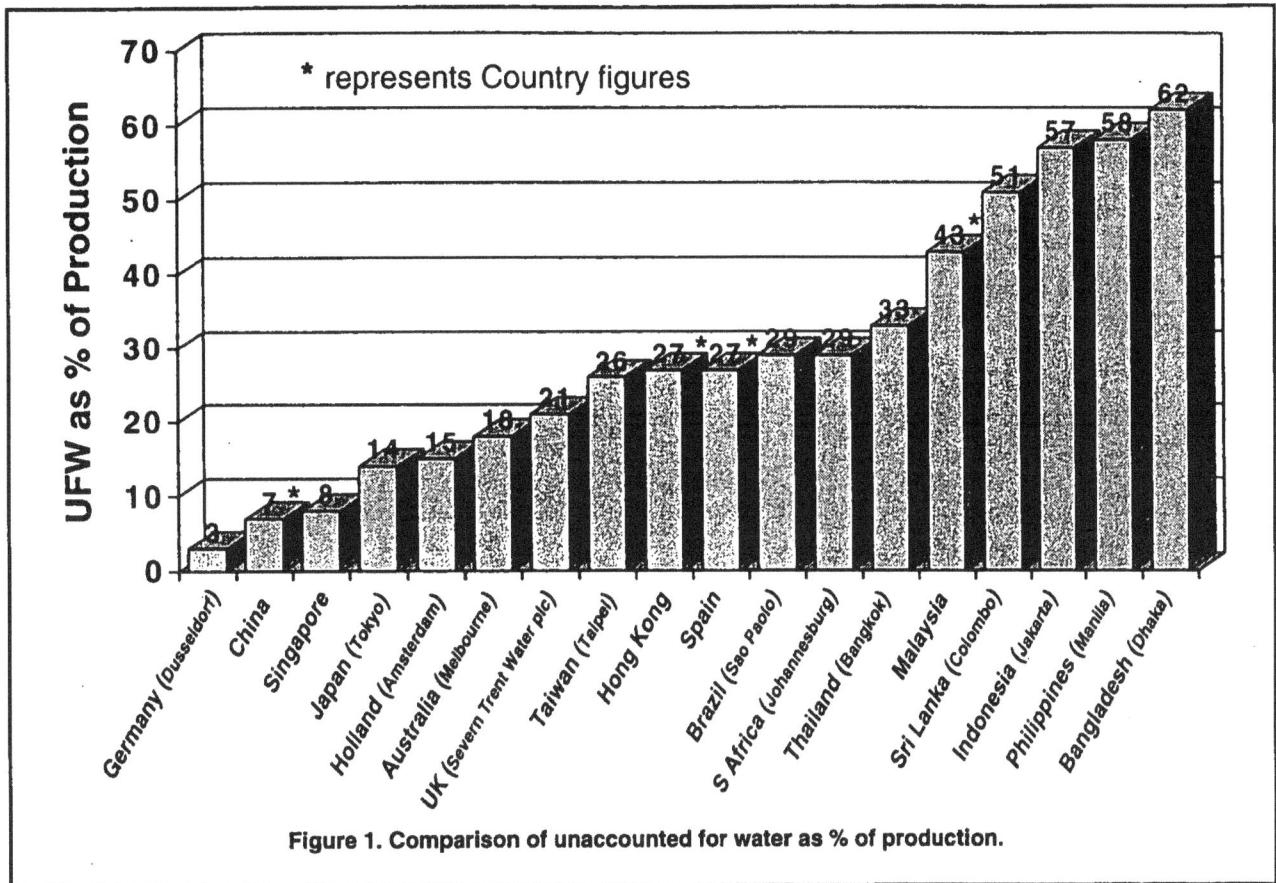

Figure 1. Comparison of unaccounted for water as % of production.

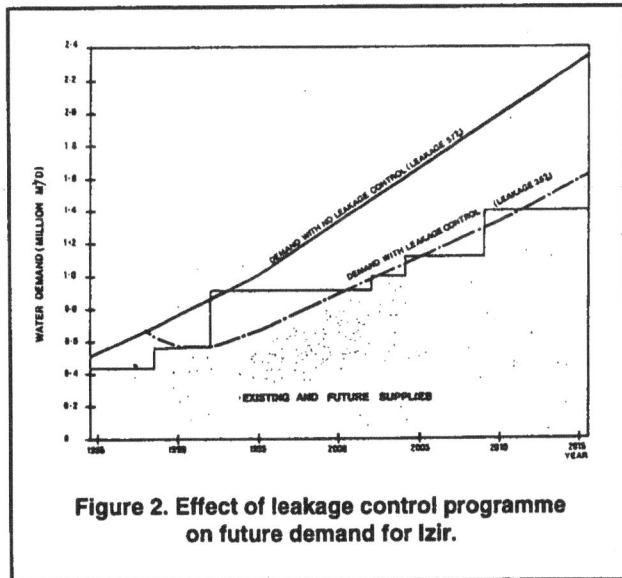

Figure 2. Effect of leakage control programme on future demand for Izir.

is tackled. In fact, just the opposite effect to that desired is likely to be achieved, as pressures will rise when the new water supply is commissioned, causing more leaks to develop and leakage from existing leaks to increase. As an example of this, when the Ngagel III plant extension was commissioned in Surabaya, Indonesia, a few years ago, it was found that less than five per cent of the additional water supplied was registered on consumer meters, the majority being lost through increased leakage (Surabaya Unaccounted Water Study, 1988). Subse-

quently, an appropriate leakage control policy was implemented, followed by initiation of a major system rehabilitation programme.

The commitment of large amounts of funding for capital works programmes demonstrates that there is a willingness, and that the financial resources are available, to tackle the real problem if only the credibility gap to promote leakage control programmes can be bridged. Furthermore, the major financial and resource investments in temporary measures, such as the tankering of supplies to affected areas and the willingness of consumers to pay for such tankered deliveries at rates well above the normal water tariff, reinforces this assertion. For instance, in Karachi there are more than 1000 tankers delivering water during daylight hours to water stress areas in which consumers may pay private contractors up to ten times the tariff rate for the water delivered. The solution being implemented now in Karachi is to progressively develop an active leakage control programme to provide short term improvements and to initiate a longer term leakage control strategy in conjunction with the development of a major new water supply system to serve the city in the medium term. This complementary approach can be very effective as leakage control policies can be implemented much quicker and cheaper than capital works programmes (these may be required anyway to meet projected shortfalls in system capacity even with leakage control).

It must be recognized, however, that although a leakage control policy may be justifiable economically, it may not be practicable to implement. This is the current situation in Karachi and Colombo, where passive control and visible leak inspections, supplemented by sounding, are being implemented despite the greater economic benefits of more intensive zonal metering policies. Until such time as service pressures and supply continuity can be improved sufficiently, then these optimal policies will remain longer term strategy goals.

UFW and leakage control programmes require a major investment in human resources as they entail fairly labour intensive activities despite the development of modern detection equipment. By necessity an intimate knowledge of the distribution system is required to operate such programmes efficiently. In addition, some activities can cause considerable disruption to consumer supplies (for example valving operations to temporarily create waste zones or to undertake step tests). Leakage control activities should normally, therefore, be undertaken by utility staff as part of routine O & M activities. A recent development has been the appointment of contractors and consultants to reduce UFW, with payment being made on results achieved. Such an approach requires care to be taken to avoid making disproportionate payments for the work undertaken, but at the same time providing incentives to encourage good performance.

Water utilities must recognize the importance of leakage control through the appointment of suitably experienced staff at the highest level of management, supported by capable staff within a permanent institutional structure. It is essential for competent and well motivated staff to be attracted into the work not only to maintain the high calibre of technical expertise and committment required but also to minimise staff turnover. In practice it is not unusual for leakage staff to be transferred from their normal posts to fill vacant posts that are deeemed to be more important within the utility. Similarly, the allocation of unsuitable staff simply to fill the required staffing complement must be avoided. The whole ethos of the UFW/Leakage Control Department, as well as that of the top management level within the utility, must be such as to encourage committment, motivation, and the determination to maintain an effective and continuous effort to control UFW and leakage. This will be reflected in the investment in personnel, training, specialist equipment, management information systems, etc.

UFW and leakage control may not be glamorous areas of work within the utility but they are extremely important nevertheless and can make a major financial impact. For instance, if the level of UFW in Colombo could be reduced to 25 per cent of production, that would be equivalent to saving up to half the capital cost of the 182 Megalitres per day Ambatale plant extension recently commissioned at a total cost of nearly Rupees 2100 million (Rupees 48.7 = US$ 1 at May 1994). Furthermore, the recurrent annual costs of treating and pumping the additional water saved would also be eliminated.

Mott MacDonald has undertaken many leakage control projects in countries as diverse as Oman, Turkey, Mozambique, Indonesia, Malaysia, Pakistan, Abu Dhabi, Hong Kong, China and currently Sri Lanka. In all of them appropriate leakage control strategies have been developed and have been proven to be cost-effective. The activities in such programmes, however, must be maintained if the improvements gained are not to be lost as a result of continuing system deterioration. Leakage control must therefore be viewed as a major component of the regular operation and maintenance activities undertaken by the water utility, with fully dedicated and well trained staff allocated to monitor, detect and control it. Funding agencies, and increasingly the utilities themselves, recognize the effectiveness and relevance of perpetual leakage control programmes, in particular where labour costs are relatively inexpensive and water is scarce or expensive to produce and distribute. Too often in the past, however, the impetus and benefits achieved from specific leakage control projects have been lost afterwards due to reduced committment from water utilities.

Major benefits are obtainable using relatively simple techniques and inexpensive equipment initially, before progressing to more sophisticated approaches. Simply locating and repairing visible leaks promptly will provide invaluable preparation for initiating a more comprehensive policy. The key to success is the committment and motivation within the utility to tackle the problem effectively, from the highest level of management downwards. The task may not command a high profile or receive public accolades, but it can be as effective as a major capital works programme. In view of the increasing demand for water and its scarcity worldwide greater emphasis must be given to leakage control and UFW programmes if we are to avoid major problems in the future. We neglect it at our peril!

References

'International Report on Unaccounted for Water and the Economics of Leak Detection', IWSA Seminar, 1991.

Lackington D W, 'Leakage Control, Reliability and Quality of Supply', *Civil Engineering Systems*, 1991, Vol 8.

'Leakage Control and Distribution Management: Lessons from Japan', Water Services Association Seminar, UK, March 1994.

'Surabaya Unaccounted Water Study, Final Report', Mott Macdonald, 1988.

'The Cost of Water Delivered and Sewage Collected 1992-1993', Office of Water Services, UK, October 1993.

'Water Malaysia 1992, 8th ASPAC-IWSA Regional Water Supply Conference and Exhibition, Technical Proceedings', Kuala Lumpur, October 1992.

'Water Utilities Data Book, Asian and Pacific Region', Asian Development Bank, Manila, November 1993.

Water and sanitation for Mahaweli

R.L. Haturusinha (Ms), K. Theivasagayam and D.T. Dáyarathne Perera, Meca, Colombo

MAHAWELI GANGA development project, the largest multipurpose river basin development project ever undertaken by Sri Lanka, involves production of 508 MW of power and development of 364 000 Ha. of land for irrigated agriculture and human settlement. The number of settler families in the entire development area would amount to about 250 000. The settlement is mostly confined to the dry zone area of the country which has a dry spell from April to September. The project area has been divided in to 13 irrigation systems from system A to M, with natural features forming the boundaries. These systems have been further divided into zones and each zone into management blocks of extent ranging from 1000-1500 Ha. for the convenience of proper irrigation and settlement management. The smallest settler unit is a hamlet consisting of 200 - 400 families. The social infrastructure facilities for the settlers are provided through service centres at different levels namely hamlet centres, village centres, area centres and town Centres. Each hamlet has a hamlet centre and each 3-4 hamlet centres are served by a village centre. Each block has an area centre and each system has one or more town Centres according to the population distribution.

Provision of safe drinking water and sanitation facilities is a key area in the quest for raising the quality of life in the Mahaweli settlements. The aim of this paper is to highlight our achievements in meeting this objective within the limits of affordability. The approach to affordable water supply has been to go for appropriate methods to suit the population distribution, the topography and the available sources of supply, keeping it economical while ensuring safety.

Each hamlet under the Mahaweli settlement consists of 200-400 farmer families each provided with 0.2 ha of homestead area. Thus the population in the hamlets is thinly spreaded. These hamlets are usually surrounded by the agricultural lands with high water table. Ground water which does not need treatment and found to be safe for consumption, forms the source for household consumption. The most affordable system of water supply for hamlets are dug wells. Each household is encouraged to construct their dug wells by providing a subsidy.

The farmers are supplied with precast concrete rings of 1 m diameter, 0.05m thickness and 0.75 m height for the construction of the wells. They carry out the construction of their own wells, by excavating up to the required depth of ground water and placing the rings up to the ground level. After the backfill a brick wall ring is constructed above the ground level up to about .75 m above the

ground level. The bottom ring is a special one with porous surface to allow seepage of water. A typical well is shown in fig. 1. The average depths of these wells are around 5-6 m which needs around 7 to 8 rings. As the rings are produced for this mass requirement in bulk at a central supply station, the cost is considerably low and is convenient, quick and easy to construct. The cost of eight units of these rings needed for an average well and a bag of cement supplied for the masonry work cover the subsidy of US$ 50 provided for each household for this purpose.

In places where the water table was deeper and below rock level the individual wells are unaffordable for the farmer, and larger community wells are provided at one per 6 families. They are about 3 m in diameter up to about 9 m in depth and with 225 mm rubble masonry wall up to 0.75 m above the ground. The cost of a community well is around US $1500. Farmers contribute to the construction by providing the labour required.

In areas where the ground water table is much deeper, tube wells fitted with simple Indian made hand pumps have been provided. Water Resources Board of Sri Lanka and a few private firms are engaged in its construction. Average depth of these tube wells are 40 m most of which is in rock strata. One or two such wells are provided per settlement where it is necessary, at appropriate locations. The overall cost of a tube wells is around US $ 2000.

The quality of the ground water is good in most of the hamlet areas and is safe for consumption. But in certain high land areas the individual wells go dry during the driest months of August-September when ground water table drops to levels deep below the rock strata. Also there are locations where the water is hard with $CaCO_3$ content. In such situations, presently there is no affordable alternative for the settlers other than surface sources such as irrigation canals and minor reservoirs, the water from which is used for consumption after boiling well.

For the bulk needs of the settlers such as washing and cattle bathing, minor reservoirs are used. They are continuously supplied by the irrigation canals and are reliable sources of water, having water round the year.

The service centres which provide social infrastructure facilities are more confined and closely packed compared to the hamlets.

The most basic one, the hamlet centre, has a few offices, a minor school, a few shops and staff quarters, have individual wells as the most appropriate and affordable system for such a small population.

The village centre which serves three or four hamlets as stated earlier, has a larger population and it is intended to serve the water supply needs of the population by a large dug well constructed at a location where the general demand of about 100 cu.m. per day could be met by ground water which does not need any treatment. The water will be pumped to a 50 m, capacity overhead tank from where it will be distributed through a PVC pipe distribution system.

The area centre which is the centre of administration for the management block holds a substantial population numbering around 3000 (including floating population) and is also a centre of minor commercial activities. As the estimated demand of 300 cu.m. per day could not be met by ground water, surface water sources are used, which need treatment. To make it affordable, a simple treatment process of biological filtration using slow sand filter units and chlorination is adopted. An overhead tank of capacity 150 cu.m. and a PVC distribution net work complete the scheme. Even though the capital cost for the scheme is around US $ 200 000 the operational cost is very low, amounting to US $ 600 per month, spent on operation of intake and clear water pumps and the wages of six unskilled labourers who are needed to maintain it.

The town centre which is the main centre of activities of the Mahaweli settlements has a population of around 10 000 on the average, including floating population. The average demand of a typical town centre such as Dehiattakandiya in System C of the Project area is around 1500 cu,m. per day. Inevitably, in such a case as that of Dehiattkandiya township, where the source of water is the Mahaweli river with widely fluctuating sediment content, a process of chemical treatment (coagulation and flocculation using alum), sedimentation, rapid sand filtration and chlorination had to be adopted. However, effort was taken to make it affordable by using an energy saving hydraulic flocculator, gravity methods for aeration, sludge removal in sedimentation tank and drainage, and using PVC pipes for raw water main and distribution net work. A tariff in line with that of National Water Supply and Drainage Board of Sri Lanka is imposed on the consumers who are mostly nonfarming population with better income to afford this tariff, to cover the operational cost.

The capital cost of the Dehiattakandiya Township Water Supply Scheme is around US $ 1.2 million and the operational cost is around US $ 2000 per month.

Sanitation facilities for hamlets too are provided on an individual basis, considering the dispersed nature of the population. A subsidy of US $ 9 is given to each household which construct a latrine. To make this possible within the subsidy provided, a design of latrine in the form shown in fig. 2 is suggested to the settlers and precast concrete slab as shown in fig. 2(a) is supplied at US $ 4 each. These slabs are precast on a large scale at a central location and hence the cost is very low and is convenient and quick for construction. The farmers themselves could do the construction and the shelter on top could be a simple one with jungle pole frames and palm leave claddings or 110 mm. thick brick work, according to the affordability of the farmer. The slabs provided earlier did not have a water seal arrangement and the opening above the pit is kept covered by a plank when not being used. This arrangement is found to be unpopular with the settlers. Now a water seal arrangement with a two piece precast concrete slab and squatting pan unit is supplied at US$ 5.5. A sketch of this arrangement is shown in fig.2(b). This is more popular with the settlers as it seals off the pit below, and could be relocated over a new pit once a pit has to be closed. There is no health hazard with these simple systems as the households have a lavish area of 0.2 ha each and the latrines could be located at a safe distances away from the water sources.

For the service centres the common arrangement is to have latrines with a septic tank and soakage pit with a capacity to suit the number of occupants for each individual household. A combined system of sewage lines and a common septic tank for several households together was thought about for the town centres but later abandoned, considering the relative advantages of indi-

Table 1: Information on wells in sample Hamlets in system C of Mahaweli Project areas								
Zone	Block	Hamlet	No of families	Permant wells	Temporary wells	% Having wells	Community wells	Tube wells
2	202	Millettewa	435	15	189	43	2	—
	204	Hembarawa	350	167	15	52	2	1
3	303	Wewagama	393	54	114	43	2	—
	304	Medagama	308	109	26	44	1	—
4	401	Sandunpura	311	55	20	24	—	—
	404	Bakmeedeniya	313	204	10	68	—	1
5	501	Suriyapokuna	230	76	20	42	1	—
	503	Nikawatalanda	255	70	20	28	1	—
6	601	Gal-Eliya	246	7	—	32	—	—
	603	Kekuluwela	267	26	3	11	—	—

NOTE-

1. THE SKETCHES ARE NOT TO SCALE.

2. DIMENSIONS ARE IN MILLI METER.

GENERAL RING

FILTER RING

RING - PLAN

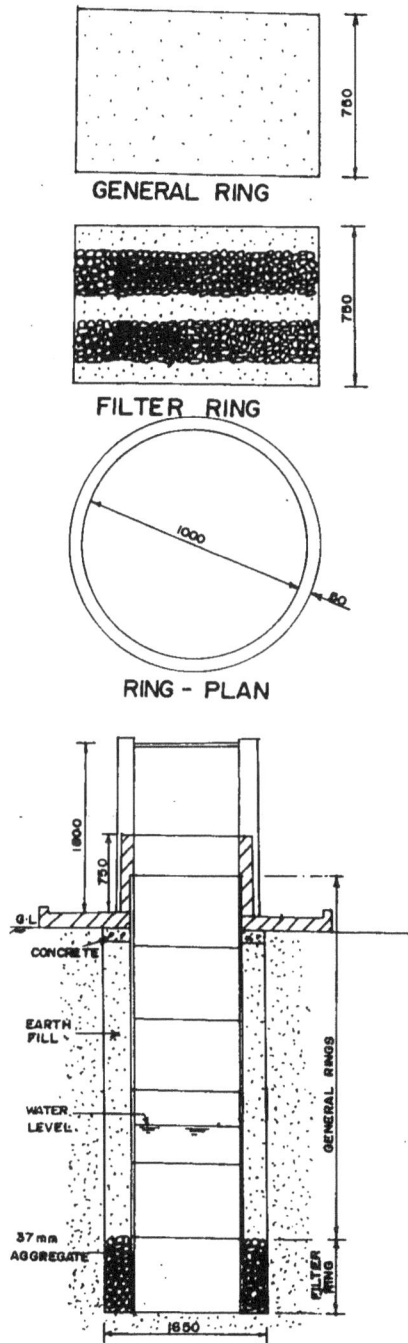

Figure 1. Dug well for settlers.

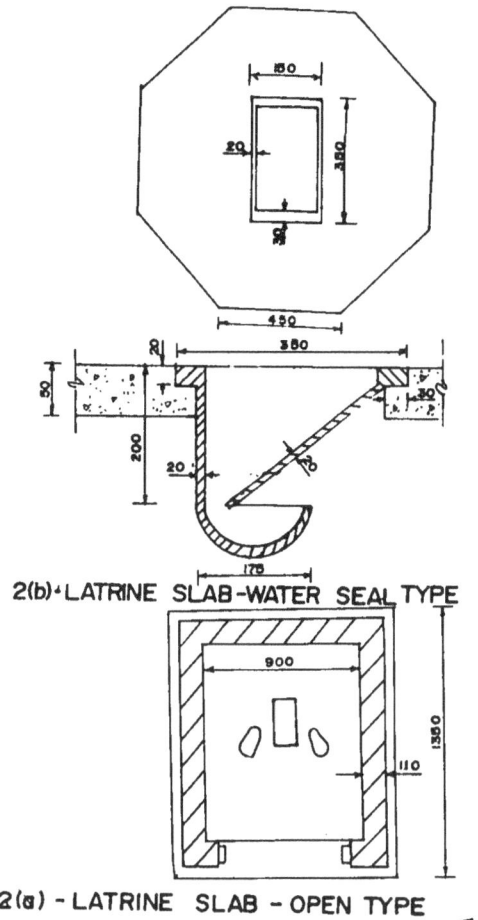

2(b) - LATRINE SLAB-WATER SEAL TYPE

2(a) - LATRINE SLAB - OPEN TYPE

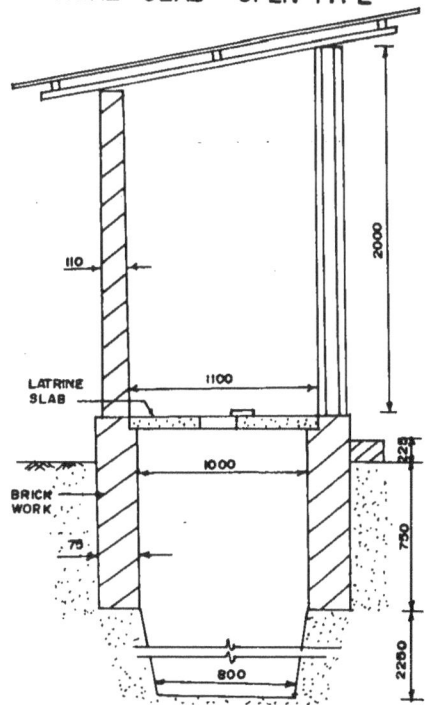

Figure 2. Latrine for settlers

134

vidual septic tanks which could be maintained easily by the respective owners.

There were special situations in certain areas where the rock level was very high and the construction of septic tank was not possible. A typical case was the Welikanda town centre in System B of the project area where the earth overburden is very shallow, varying from 1 to 2m in the entire area. Hence an integrated sewerage system consisting of a pipe network leading into a stabilization pond with a detention period of one month located away from the township was found to be an appropriate solution. As the entire system was designed to work on gravity flow the operational cost is negligible.

A survey was done by the authors of the paper with the assistance of the staff of the Resident Project Manager, Mahaweli Economic Agency in System C of the project area which could be considered as a representative sample for the Mahaweli Systems, to review the situation of the water supply arrangements specifically in the settlements (Hamlets). Two representative hamlets in each zone from zone 2 to 6 was surveyed and the information obtained is given in table I.

During our survey and interviews with the settlers and the project staff, we were able to observe the following:

- With the decrease in subsidies for individual dug wells from the earlier value of US$ 50 to the present US $ 20 the percentage of individual dug wells constructed have significantly dropped. (see from Table I as in the case of recently completed zones 5 and 6 where the subsidies have been reduced)
- In the case of many individual wells, excavation was stopped at a level when they encountered rock. Most of these wells go dry during the driest spell in August - September. In such cases if the settlers are helped in blasting the rock to sufficient depth, it would ensure continuous supply of water.
- In certain areas the ground water contains salts causing hardness, colour and taste. Investigations have not been done to ascertain the ground water quality and depth of water table in the settlement areas before deciding on these locations.
- The progress in construction of tube wells, which would have been a boost to the settlers during the period when most of the individual wells go dry, lags very much behind the schedule due to lack of funds

for this purpose. This programme does not get much priority.
- In certain cases, there is lack of response from the settlers for community wells even though funds are available for this purpose. This may be due to lack of group effort.

Based on the above observations we would like to make the following recommendations for consideration in future human settlement projects of this nature.

- Provision of safe drinking water and sanitary facilities must be a priority item in the human settlement projects. The development agencies must either provide these facilities or encourage settlers to do it by providing subsidies up to realistic levels.
- In the planning stages, investigations regarding ground water quality and depth too has be done and the results be taken into consideration in identifying settlement areas, except in cases where alternatives for affordable and safe water for consumption are available.
- In settlements where the ground water table fluctuates heavily so that the dug wells dry up for a period, tube wells have to be provided to ensure availability of safe drinking water during dry periods as well. The aid programmes for safe water and sanitation should cover tube-wells too.
- Assistance should be provided to settlers for rock blasting in the process of excavating the dug wells in cases where it will ensure continued supply of water at reasonable depth.

References

[1] Mahaweli Ganga Irrigation and Hydro-Power Survey, Ceylon, Volume 1: General Report, Phase 1; Ad Hoc Report, UNDP/FAO.

[2] Instruction notes to settlers for construction of Dug Wells and Latrines: Mahaweli Economic Agency of Mahaweli Authority of Sri Lanka.

[3] Type plans, Estimates and construction records for water supply and sanitation works for settlers, kept at the Resident Project Manager's Office (System C), Mahaweli Economic Agency of Mahaweli Authority of Sri Lanka.

Spill-water recycling

Mrs. N. Kamalamma, E.K.N. Varma Raja, Mrs. N. Prema and Miss K.S. Pushpa, India

ALL OUR LABORATORIES are devil's paradise unless they serve the poor. Gandhiji.

In order to serve the poor, our laboratories must generate innovative technologies or modify the existing ones to suit location-specific requirements and transfer the new technologies to rural areas. This paper presents one such effort made to tide over the twin problems of water stagnation around the community source on the one hand and shortage of water on the other.

Waste-water stagnation around a community water source is caused primarily by the spill-over from water pitchers. A rural family around Gandhigram collects 8 to 20 pitchers of water every day (1992). On an average, while filling a pitcher, two litres of water gets spilled. Accordingly, at a source serving 50 families, around 800 to 2000 litres of water gets spilled. Day in and day out, when the activity is continued, this water accumulates, creating serious environmental problems — water stagnation around the source leading to contamination of water at source and mosquito breeding.

Figure 1. Spill-water stagnating around a handpump

The spill-water is not heavily contaminated; it contains primarily the mud and the dirt on the feet of the water carrier. Therefore this water can be easily and inexpensively purified and made available for uses other than drinking and cooking, thereby effecting considerable conservation of water resources.

Principle

The technology of recycling spill-water is very simple. If the spill-water, after separation of bigger-sized foreign materials, is allowed to move slowly through a filter column of graded granite chips, a major portion of the solid impurities would get trapped in the filter medium by gravitational force and statical blockage and the water that is let out would be clear and usable.

Since bacteria have greater attraction towards silt and dirt rather than to water, they will also be held up in the filter bed thus resulting in improvement in water quality.

Procedure

Figures 2 and 3 show a spill-water recycling unit at a water source.

Three units with provision for inlet and outlet slots form the system for recycling the waste-water. The first unit is designed to remove solids and silt; the second unit facilitates filtration by forced flow through graded granite chips placed in order; and, the third unit collects the filtered water. The supernatent water from the third unit can be taken by a pipe-line to a convenient place for storage and use.

Construction

The construction of a spill-water recycling unit is simple. The device consists largely of simple masonry structures which can be prefabricated. The precasting of the units can be done using very simple techniques. Ferro-cement skeletons of the units can be fabricated using mild steel rods/wires, preferably welded or tied together with wires to the desired dimensions. Chicken-wire-mesh is spread over the structure so as to retain the cement plaster in shape. This moulded structure has to undergo the usual watering, curing etc., before it is ready to be installed.

The unit can be installed at convenient locations considering the elevation and slope required for steady and slow movement of water. By adjusting the length and breadth of the filter column the technology can be further adapted to locational requirements.

The units need to be cleaned and recharged periodically. The methods are very simple, like back-washing and disposal of muddy water. The leach pit connected to the first unit facilitates easy disposal of muddy water through absorption by surrounding soil.

Effect of recycling

Eighty per cent of the water wasted at a source can be easily realised for re-use with this technology.

Effect of recycling on water quality

Parametres	Before recycling	After recycling	Reduction of impurities in %
Bacterial count (MPN/100ml)	1609×10^4	240×10^4	85.08
Total solids (gms/100ml)	0.0245	0.0125	48.97

The filter medium acts both as a mechanical barrier and as a biological one and facilitates purification of the water fed into the system.

A spill-water recycling unit is basically a community project. So, before installing a unit, the readiness of the community to maintain the system through regular cleaning is to be ascertained. One way of ensuring regular maintenance of the spill-water recycling unit is to enjoin it on the personnel employed under the social forestry programme or similar economic programmes requiring water.

Several communities around Gandhigram have accepted the concept and technology of spill-water recycling and are using the recycled water for laundering and pisciculture. In an institution at Gandhigram recycled spill-water from a single hand-pump has been used for watering trees and about 300 saplings could be nurtured in this dry area where, normally, water had to be procured from outside even for essential requirements. Several other organizations have also sought assistance in installing spill-water recycling units.

Conclusion:

Since water which would otherwise have been wasted is conserved and reused, the technology becomes relevant and meaningful for environmental upgradation.

Reference

1. Kamalamma, N. and WVarma Raja, E.K.N. (1992). Spill-Water Recycling. Gandhigram Rural Institute, Gandhigram.

2. Kamalamma, N. (1993). Soakpit and Spill-water Technology. Gandhigram Rural Institute, Gandhigram.

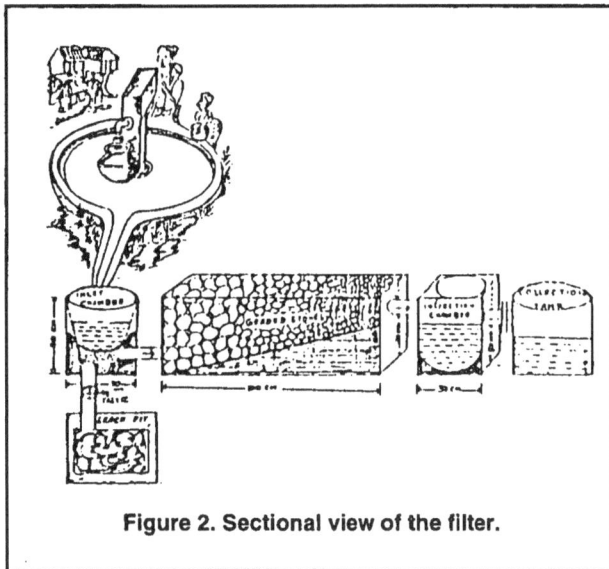

Figure 2. Sectional view of the filter.

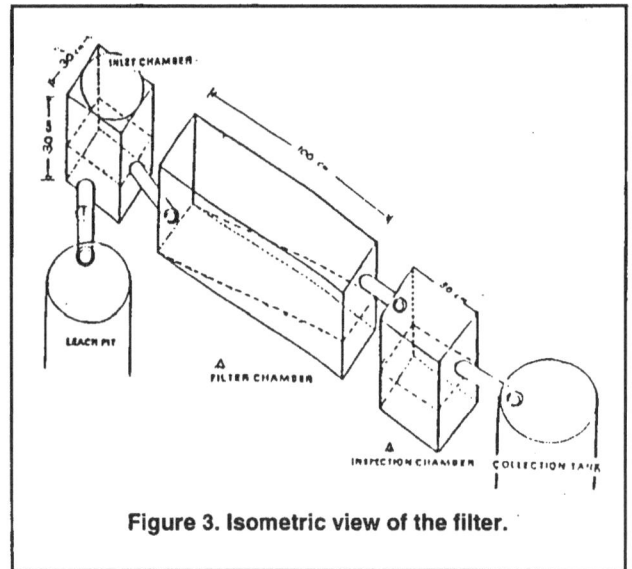

Figure 3. Isometric view of the filter.

Water problems in the Jaffna Peninsula

Dr. V. Navaratnarajah, University of Jaffna, Sri Lanka

THE POPULATION IN the Jaffna Peninsula has always depended on the water stored in the underground miocene limestone and sand aquifers for its drinking water and water for irrigation to agricultural lands. Although there is a small acreage of paddy lands, the majority of the cultivated lands is used for agricultural activity related to short term crops. Water for drinking and for agriculture is obtained from open wells. Even the municipal area in Jaffna is supplied with water pumped from wells located in Kondavil, about three miles from Jaffna Town. Open wells are commonly built in villages, one for each household albeit with some of them shared by three or four neighbouring households who occupants are usually related to one and another. A recent innovation is the sinking of tube-wells to obtain water supply for drinking and irrigation to agricultural lands.

Quality of water and affordable water supply

In the early days, well sweeps or a system of pulleys were used to extract the water from the open wells, both for consumption as drinking water and for irrigation to crops. The fresh water in the aquifers floats in lens formation or varying thickness on saline water found below and has salinity levels depending on its location and distance from the sea. For example, a study on the geomorphology of the Valukkai Aru drainage basin in the Valigamam area showed that the salinity of the ground water in a location is inversely related to its distance from the sea. (Puvaneswaran, 1987). The amount of calcium, magnesium bicarbonates and sulphates present in the water contributed to its hardness which varied from moderately hard to hard in terms of calcium carbonate equivalent. Since the rate of extraction of the underground water was relatively slow, the recharge of the wells from the underground reservoir helped to maintain an affordable water supply of satisfactory quality in terms of salinity and hardness.

Ground water recharge has been viewed as a function of effective rainfall. In the Jaffna Peninsula, this occurs only during the annual monsoon rainfall during the period September to January. After losses by direct runoff (about 10-15 per cent), and losses by evaporation (about 40-48 per cent), only 30 to 32 per cent of the rainfall is left over for ground water recharge. In the last three decades, the quality of water in the region has deteriorated due to various reasons. Variability of the rainfall over the region has indirectly contributed to this as rainfall over the region has indirectly contributed to this as rainfall is the only source of recharge. A sample of rainfall records for the period 1985-1993 given in Table 1 supports the variability of seasonal rainfall. The salinity problem was perceived as a hazard as early as the 1950's and 1960's and this has been attributed to the dry periods identified during these two decades (Puvaneswaran, 1985).

A significant factor that has contributed to increased salinity in the well water has been the indiscriminate extraction of water from the underground aquifers. This has been exacerbated by the increase in population in the region and the rapid rate of extraction using pumps, both electric and petrol driven for domestic and agricultural purposes. Because of low fresh water heads in the underground aquifers, large amount of withdrawals from wells cause heads to decline further and the fresh water-salt water interface to rise resulting in salt intrusions in several areas in the peninsula (Nandakumar, 1983). The intensive agricultural pattern adopted in the last three decades also led to the increase of salinity in the water. Several wells once used to supply potable water are not in use now due to the increase of salinity (Nandakumar, 1983).

Another factor that has surfaced is the high level of nitrates and nitrites in the drinking water in the peninsula. It is well known that nitrates and nitrites above a certain level in drinking water and soil may cause serious health problems due to their toxicity. It has been reported that if the drinking water contains more than 10 ppm nitrate-nitrogen (45 ppm nitrate), it could affect the health of infants, giving rise to blue babies (WHO, 1971). The high level of nitrates and nitrites in the soil and water in the peninsula is attributed to the abundant and indiscriminate use of chemical fertilizers, mainly urea which contains 46 per cent nitrogen. The problem has been further accentuated by the improper planning of soakage pits and latrines which leads to serious contamination of the ground water by nitrates.

The recent study of nitrates in the soil and well water (Mageswaran and Mahalingam, 1983) has shown that in several areas in the Jaffna Peninsula, the soil samples have nitrates above the safe level.

In places where there is no agricultural activity, the amount of nitrates in the soil adjoining the well is below 20 ppm in majority of cases, whereas in areas where there is cultivation, the soil adjoining the well seems to have fairly large quantity of nitrates, usually above 30 ppm. The water samples from wells in areas where there is no cultivation such as Jaffna Town, Kopay, Kokuvil, Uduvil,

Nallur, Tellippalai, Mirusivil, Naranthanai and Karaveddy have nitrate - nitrogen less than 18 ppm. The water samples in wells in plots where there is agricultural activity have nitrate - nitrogen levels between 20 to 50 ppm. The villages Kondavil and Urumpirai where there is intense agricultural activity have very high nitrate - nitrogen levels of 30 ppm.

The water samples from the wells in Thirunelvely and Kondavil form which water is drawn from Jaffna Town supply have a high nitrate - nitrogen level of 26 - 33 ppm which is about three times the safe level. The investigations (Mageswaran and Mahalingam, 1983) also indicate that the nitrate - nitrogen levels increase year by year. For example, the nitrate - nitrogen level of Thirunelvely well water increased from 15 ppm in December 1976 to about 22 ppm in December 1980 and to about 27 ppm in May 1982. Similarly, the water in the wells in Kondavil increased from 22 ppm in December 1976 to about 30 ppm in December 1980 and to about 34 ppm in May 1982.

The above study also showed that in parts of Jaffna Town where there is no cultivation, the nitrate - nitrogen level in the well water approaches 20 ppm. This is probably due to the wells in thickly populated areas being closely situated to the soakage pits of toilets. With increasing demand for houses, the local authorities are willing to reduce the maximum distance between wells and septic tanks from 15 to 5 metres. This could case serious health problems, as the limestone rock is fairly close to the ground surface and sufficient depth of soil soakage, absorption and filtering is not available. A study made of the salinity and nitrate values in the soil and water in selected areas in 1992 (Baskaran, 1992) showed very much more increased values in the soil and water which are of concern to the health authorities. A representative sample of the values is given in Table 2 and 3.

It is evident that in discussing affordable water supply, aspects such as ensuring that the quality of the water supply meets the required health standards and cost involved in doing so have to be considered.

The economics of affordable water supply

In relevance to the discussion on water quality, in order to ensure that the quality of water in the Jaffna Peninsula is brought up to a satisfactory level and maintained at the level at least in terms of salinity and the nitrate - nitrogen levels, various measures have to be adopted. Amongst them are the following:

o Satisfactory systems of wastewater disposal for the Jaffna town municipal area and other urban areas where the households per acreage is high.

• Promote the use of biofertilizers instead of using chemical fertilizers in agriculture.

• Increase the recharge to the underground fresh water reservoirs in the peninsula.

A satisfactory system of pipe-borne wastewater disposal system for the municipal and urban areas would also involve pipe-borne water supply systems and the related purification and treatment systems. Although plans have been drawn for the installation of such systems in the Jaffna municipality and urban areas such as, for example, Chavakachcheri, these have not been implemented due to various reasons, an important one of which is the lack of funds. Eventually, when these systems are installed, the cost of these undoubtedly would be passed on to the consumers in the form of higher urban taxes.

In the rural areas and areas outside the higher density housing areas and in particular where there is agricultural activity, steps could be initiated to reduce the Nitrate - nitrogen levels in the well water by reverting back to biofertilizers. Before the introduction of chemical fertilizers, the farmers had been using biofertilizers such as biomass (leaves of trees) and cattle waste products, and during that period, the quality of water has remained satisfactory and did not warrant any further treatment before use. Hence, the use of biofertilizers would remove one of principal factors of the increased nitrate - nitrogen level in well water.

It has been shown (Puvaneswaran, 1985) that the salinity of water in underground reservoirs increased when the recharge from the rainfall was reduced. Hence, steps should be taken to increase the recharge to the underground resources by conserving more of the rainwater. In the early days, there existed a large number of ponds and tanks with interconnected channels that helped to conserve a large proportion of the rain water and also to recharge the underground resources by conserving more of the rain water. In the early days, there existed a large number of ponds and tanks with interconnected channels that helped to conserve a large proportion of the rain water and also to recharge the underground reservoirs. However, over a period of time, particularly during the last three or four decades, with increased pressure for housing, several of these ponds and tanks have disappeared, with development on these locations after the tank and ponds have been filled. At the present moment, the rehabilitation of the remaining tanks and ponds is being actively undertaken.

In order to meet the water needs of the Jaffna Peninsula, a scheme was launched in 1952 by the then Irrigation Department to convert gradually the water in the lagoons in the Peninsula to fresh water by flushing out the salt in the natural manner with the monsoon rains over a period of time. The scheme was almost completed in early 1970's when water in the lagoons had been recognised to be fresh water and the farmers cultivating the lands adjoining the lagoons were deriving the benefits. It was also reported that the wells near the lagoons which had showed salinity earlier were now fresh water wells providing water suitable for drinking and agricultural purposes. Unfortunately, due to poor maintenance of the scheme and various other factors, the lagoon scheme had fallen into a deteriorated condition with the water in the lagoon becoming saline and unfit for use. The main advantage of converting the water in the lagoons into fresh water is that it helps to recharge the underground reservoir with fresh water. As a result, the adjoining lands become rid of salinity and suitable for agricultural purposes. In addition, the wells in the area also become less saline.

A recent study to rehabilitate the Jaffna Lagoon Scheme and the tanks in the Peninsula suggested that this can be done at an estimated cost of SLR 300 million (US$ 6 million). In return, the benefit in terms of increased agriculture, increased fodder from grass grown in adjoining lands and the improved health of the population that would result from the consumption of the increased production of milk and milk products was estimated at SLR 152 million (US$ 3 million) a year, suggesting that the cost would be returned in two years. However, what is more important is that the scheme once in operation would continue to provide the Jaffna Peninsula with fresh water.

In conclusion, in order to provide an affordable satisfactory quality water to the population in the Jaffna Peninsula, perhaps the first priority seems to be the rehabilitation of the Jaffna Lagoon Scheme. This could be followed by the institution of pipe borne water supply and waste disposal systems as funds become available.

References

[1] Baskaran S., 'Analysis of water and soil samples from selected areas in the Jaffna Peninsula', *Graduation Report of Special Degree in Chemistry*, University of Jaffna 1992.

[2] Mageswaran R. and Mahalingam S., 'Nitrate - nitrogen content of well water and soil from selected areas in the Jaffna Peninsula', *Journal of National Scientific Council*, Sri Lanka, 1983, V.11, No. 1, pp. 269 - 275.

[3] Nanadakumar V., 'Natural environment and ground water in the Jaffna Peninsula, Sri Lanka', *Climatological Notes* No. 33, Tsukuba, Japan, 1983, pp. 155 - 164.

[4] Puvaneswaran K.M., 'Spatial temporal variation and the human dimension of the ground water of Jaffna Peninsula', *Beitrage Zur Hydrologie*, 1985, V.5, No. 2, pp. 827 - 845.

[5] Puvaneswaran P., 'Geomorphology of the Valukkai Aru drainage basin', *Sri Lankan Journal of South Asian Studies*, 1987, V.1, pp. 43-58.

[6] 'The rehabilitation of the Jaffna Lagoon Scheme', Report submitted to the Government Agent, Jaffna, 1992 (unpublished).

[7] World Health Organisation, "International Standard for Drinking Water", 1971.

Table 1 : Rainfall records in mm in period October - December

1985 : 1190.4	1989 : 630.7	1992 : 278.2
1986 : 881.3	1990 : 1041.2	1993 : 1125.3
1987 : 568.3	1991 : 539.2	(up to 15 Dec.)

Table 2 : Salinity and Nitrate values in water in Jaffna peninsula

Location	Chloride ppm	Nitrate ppm	Ca ion ppm	Mg ion ppm	Na ion ppm	K ion ppm
Kokuvil	300	30	66	18	23	8
Jaffna Fort	540	80	73	37	28	16
Valvettithurai	166	198	26	13	14	25
Vaddukkoddai	90	43	30	13	15	6
Kurunagar	695	160	40	26	15	3

Table 3 : Salinity and Nitrate values in soil in the Jaffna peninsula

Location	Chloride ppm	Nitrate ppm	Ca ion ppm	Mg ion ppm	Na ion ppm	K ion ppm
Kokuvil	159	1200	10775	478	30	21
Jaffna Fort	312	998	7899	90	48	30
Valvettithurai	362	505	9519	451	62	40
Vaddukkoddai	61	1135	6131	132	20	42
Kurunagar	302	586	11227	160	72	46

Water pumping technologies - NERD experience

Dr A.W.S. Kulasinghe, Chairman, National Engineering Research and Development Centre, Sri Lanka

DEVELOPMENT OF SUITABLE pumping devices to meet the growing demand for water pumping has been a great challenge for those involved in research and development work in that discipline. The variables governing the successful operation of a water pump are numerous and they become particularly difficult to control in the third world developing countries.

Suitability of a pumping device depend on many factors. Availability of primary energy sources which could provide drive power for pumps stands atop of them. In fact many of the conventional pumping stands atop of them. In fact many of the conventional pumping systems which require some form of commercial energy for its operation are unsuitable for places where there is inadequate supply of commercial energies — for example non availability of grid power, unaffordability of fuel oils due to poor economic standing of consumers, etc.

Other factors such as acceptability, maintainability, durability, affordability, etc. come next. The degree of their influence on performance and sustainability of the system can vary depending on the circumstances — culture, beliefs, education, adaptability, skills, etc. The problem becomes further complicated when trying to popularise non-conventional technologies in rural areas due to socio-economic and political issues due to comparatively lower educational standard and skill levels of consumers in these areas. Introduction of any new pumping device (pumps other than those operated on conventional energies) even with VLOM (Village Level Operation and Maintenance) features can have resistance from prospective consumers mainly due to the unfamiliarity with the new system and perceived socio-economic differences.

The situation is most demanding and forceful enough to draw the attention of the researcher to concentrate more on appropriate technologies when developing new pumping systems. National Engineering and Research and Development Centre realizing the need for different but effective pumping systems undertook some pioneering research work in developing and testing a number of non-conventional pumping devices. Some of them are based on novel concepts and others tried to improve on existing systems. The key factors taken into consideration when developing these new systems are use of non-conventional energy sources, flexibility of the system, minimum maintenance, affordability and use of local raw materials. This paper discusses some of the work undertaken by the NERD Centre in developing two types of non-conventional pumping systems, their popularization problems and other related issues.

New concepts and development of pumping devices

The NERD Centre has undertaken extensive research in the development of following non conventional pumping devices and obtained some encouraging and useful results. The most significant feature of these pumping systems is that they rely on non-conventional energy sources for their operation.

Pneumatically operated water pump

Figure 1 shows a schematic view of this pump. It operates on a new concept based on buoyancy effect to activates the pump. Compressed air provides energy for pumping. The device is automatic and intermittent in operation. Methods are available to minimize the non-pumping time, in which case pumping can be almost continuous.

Main features and advantages of the pump

- Compressed air supplies the motive power required to operate the pump. Not dependent on conventional energies.

- It is a positive displacement pump.

- Greater flexibility in the installation of pump. For example, unlike in the conventional pumps where the pump and the prime mover (electric motor or engine) are in one unit or are installed very close to each other, in this pump the compressed air generator can be installed away from the pump. This flexibility gives a lot of advantages when selecting locations for installing pumps in remote areas and when security could be a problem.

- Source of compressed air supply is immaterial — It could be either electrically driven, windmill driven, hand operated or any other.

- Since the pump works automatically it starts operating whenever the compressed air is available and stops when compressed air is cut off. When the pump is driven by a wind mill air compressor (NERD has developed this) it takes advantage of this feature, because of the windmill operation may not be regular.

- The pump can be easily fabricated out of locally available raw materials.

- The pump is submerged below the water surface, therefore, it has no suction head.

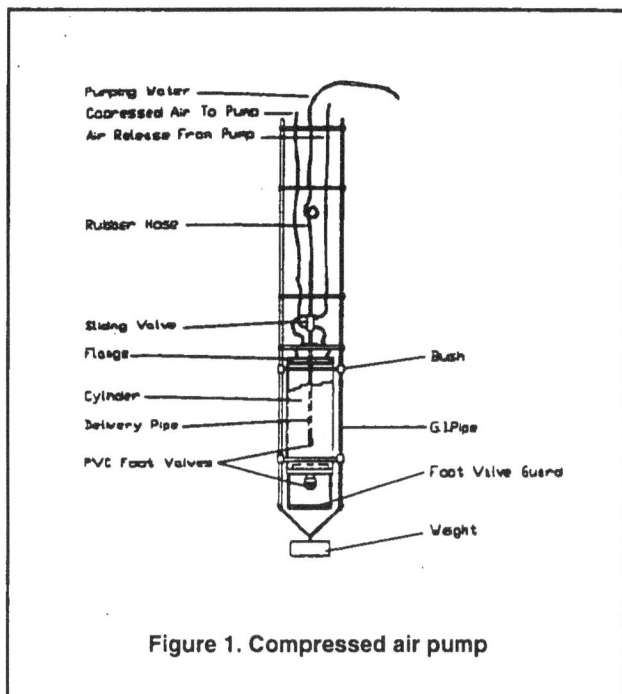

Figure 1. Compressed air pump

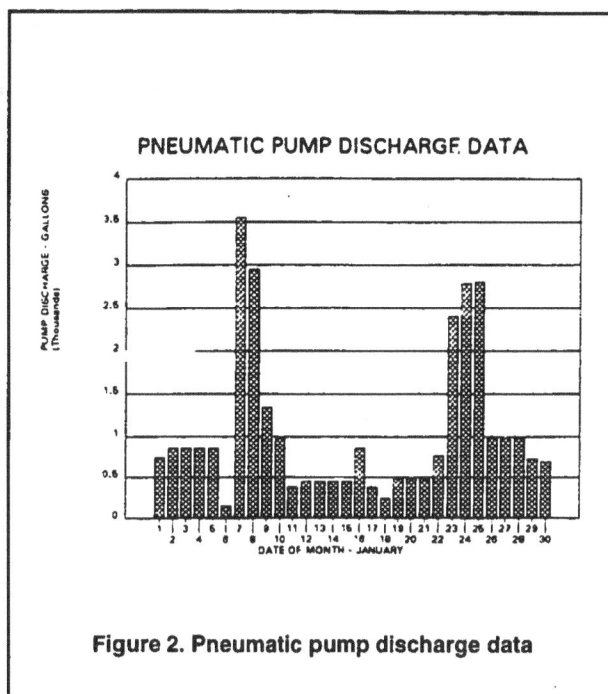

Figure 2. Pneumatic pump discharge data

Operation of the pump

When the pump (Figure 1) is lowered into water, water enters the cylinder through the bottom foot valve. Due to the weight of water in the cylinder, the cylinder moves down along the two guides. Since the valve operating mechanism is attached to the cylinder (and some parts to the guides) this movement activates the value and connects up the air supply to the pump cylinder. The compressed air then pushes out the water in the cylinder through the discharge valve. When water in the cylinder is emptied the buoyancy force acting on the cylinder pushes it up making the air valve to operate again: this time it connects the cylinder (now filled with compressed air) to atmosphere allowing compressed air in the cylinder to release and at the same time cutting off the compressed air supply to the cylinder. This action allows the cylinder to fill again with water. The operation of the pump starts again and continues in a cyclic manner. The weight connected to the guides keep the pump stable in water.

Popularization efforts

Extensive field testing carried out in the operation of the pump indicated very promising results. However, major economical and political forces tended to paralyze the popularization efforts undertaken even with the State patronage.

In 1989, for example, NERD Centre undertook a major water pumping project at Mahakanadarawa, about 150 km North of Colombo, in Anuradhapura district. The objective was to provide water for Chena cultivation for farmers in this village. Total of 11 wind mills and pneumatically operated pumping units were installed to supply nearly 20 acres of Chena land. Water has to be lifted up from open wells to a height of 4 m (from ground level to water level in the well). The Ministry of Energy Conservation through its Energy Conservation Fund arranged all funding and the North-Central Provincial Council, under whose purview the administration of the district comes, co-ordinated the implementation of the project. Initially all parties agreed to install compressed air pumps together with matching windmills to supply compressed air. So NERD Centre started the project with confidence, backed by the State guarantee, and worked hard to see that the project was completed successfully. Half way into the project, however, things began turning against the very objective of the project only to satisfy short term benefits of individuals. To our dismay the Provincial Council had arranged to distribute kerosene powered portable pumping sets to all those farmers who were earmarked for windmill driven compressed air pumps. This made people lose interest in the new pumping device even before its installation could be completed, because kerosene pumps are more flexible in usage and gave other personal advantages to the users. For example, people can hire out the kerosene pump to outsiders and generate additional income even at the expense of their own use. In spite of all these demotivating events that took place during erection of wind mills and pumps the NERD Centre managed to complete the installation and commissioning in just three and a half months. Thereafter, the system continued to operate for over one year without much problems and NERD Centre carried out whatever the routine maintenance work required during the period. In fact the project became popular among some farmers mainly due to the greatly reduced operating cost - free energy. Problems cropped up, however, as to the continued maintenance of the system after NERD Centre withdrew from the project. Initially it was agreed to hand over the project to a Pradeshiys Saba which came under the North Central Provincial council. But, training of personnel to man the project (just two craftsmen to attend to routine maintenance and other occasional repairs and adjustments work) as proposed by NERD Centre did not materialize due to various reasons which were beyond NERD's control. Issues such as funding for training, payments for work attended to, etc. did not receive concerted effort from the parties benefited by the project.

Performance

Since the pump is of positive displacement type the discharge characteristics are more or less linear and are determined by the pressure and the volume of compressed air supply. Delivery head is proportional to the air pressure and the flow to the volume of air. Figure 2 shows pump discharge when connected up to a windmill air compressor at NERD's test site.

Barrel pump — rotating coil pump

Figure 3 shows a schematic view of the barrel pump. It is a classic example of the use of non-conventional energy for pumping water to moderate heights. It utilizes the energy of a flowing stream and no regulation is normally required. It can be easily fabricated from locally available raw materials and no special skills are required.

Main features and advantages of the pump

* No external energy is required. The pump extracts energy from the flowing stream in which the pump is half immersed. Hence, the device is very useful in areas where conventional energies are in short supply or hard to get at. Sustainability is reinforced due to this.
* The pump is of very simple design and construction, and since it operates by itself when lowered into a flowing stream no operators or training of personnel

are required to operate the pump, a very positive point for sustainability of the system.

Construction

A flexible hose of predetermined diameter is wound around a water tight barrel, the diameter and length of which must be able to exert sufficient up thrust on the pump so that the barrel will float half immersed when lowered into a flowing stream. One end of the hose is open and will come in contact with water intermittently when rotating. The other end is taken out along the axis of rotation of the barrel and through a rotating joint and serves as the delivery. A number of vanes made of sheet metal are fixed to the barrel radially, in which case the axis of rotation will be perpendicular to the direction of flow of the stream. It is also possible to have the axis of rotation in the direction of flow in which case the vanes would be of propeller type fixed at one end.

Operation

Lower the assembled pump into a flowing stream — there should be sufficient depth of water to enable the pump to float freely — and tie it to the two banks with ropes or other convenient arrangement to prevent the pump from drifting down stream. If the pump is released now, it will start rotating due to the flow and water will be pumped out through the delivery line. Stream velocity of 2 to 3 ft/sec would be sufficient to operate the pump and the discharge would vary according to the size of the hose, rotational speed of drum, size of hose and the delivery head.

Test results

Table 1 gives the results of a test carried out with this pump. Details of the pump are as follows.

Hose diameter	=	2.25"
Number of coils around drum	=	12,
Drum diameter	=	25"
Number of water inlets	=	1.

Conclusion

The two pumps described above are ideally suited to conditions prevailing in many developing countries. They will also make a significant contribution to conservation of energy without creating any environmental problems.

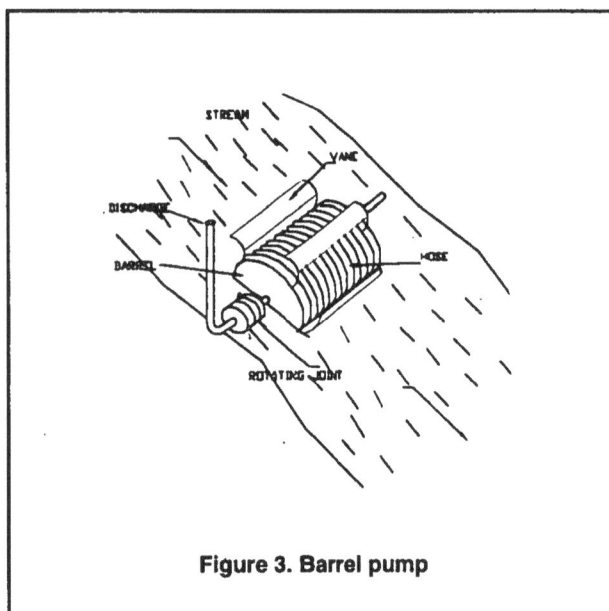

Figure 3. Barrel pump

Table 1: Rotating coil pump		
Barrel speed rpm	Delivery head - ft	Flow rate - gpm
6	8	160
7	8	180
8	8	240
7	6	148
8	6	216
10	6	270
12	6	294

Leakage and demand management

G.R.J. Warder, W.S. Atkins Consultants Ltd., UK

WATER HAS LONG been thought of in many countries as a plentiful resource. In other countries it has always been scarce but wherever, increasing demands for potable water plus increasing demands from agriculture and industry mean that there is now more competition for existing water resources. Unlimited cheap supplies of clean water are no longer possible in many countries, and significant planning of how resources should be managed to achieve an equitable and economical distribution of the water is required.

As demands for water rise to meet the capacity of existing sources, pressure to exploit the remaining available resources mounts. Many options may arise for meeting demands. These options may include:

- new source works
- transfers from regions with water to regions without
- new storage schemes
- management of demands to reduce the pressure on resources
- conjunctive use of surface and groundwater schemes
- re-designation of planning areas within catchments to allow different qualities of water to be used
- control of leakage
- more unconventional options (such as icebergs)

These options, when considered with the legislation, planning, organizational and economic implications, can form a Strategic Planning of Water Resources study. Such a plan will avoid haphazard and piecemeal development of sources, avoiding lack of control, inefficiency and over utilisation which may result in environmental damage. In one large country in the southern hemisphere, academics have recently come to the conclusion that poor planning of their national water resources is leading to severe restrictions in economic growth.

So how much effort needs to be spent on reducing leakage and on other forms of demand management and what is the validity of a demand management programme, within a wide ranging strategic resource strategy?

Demand management techniques include water saving technology, economic incentives, regulations and consumer education. They can be split into two categories; intermittent and continuous.

In the intermittent category are those measures which can be introduced when the normal demand/supply relationship goes out of balance. These include:

- Hosepipe and sprinkler bans
- Drought order

- Campaigns for voluntary savings
- Rota cuts
- Use of standpipes

These measures have increasing effectiveness. In the UK, past calls for voluntary savings have obtained reductions in demand of the order of 15 per cent.

The Director General has now made each water company agree a Level of Service Indicator, which sets return periods for these measures. Thus their use is restrained.

Continuous demand management measures include:

- Metering
- Tariff structures
- Water using appliances
- Consumer education
- Leakage and waste prevention
- Pressure reduction

Metering

Metering of domestic consumption together with an appropriate tariff is an effective way of managing demand, since it imposes an economic incentive on the consumers to reduce wastage, leaks and uneconomic use of water.

The UK National Metering Trials project was set up in April 1989 to provide information on the costs and benefits of widespread domestic metering. The Metering Trials Final Report (Ref. 1), recently published, gives the results from the running of the trials over a three year period covering:-

- Installation costs
- Operating costs
- Customer acceptability and effect on bills
- Meter location
- Effect on demand
- Technology

The trials covered 60 000 households in twelve trial areas in England, one of which, the Isle of Wight, was a large scale trial involving 50 000 properties.

The trials have confirmed that domestic metering can influence the amount of water that customers use. In the small scale trials, the average reduction in use was 11 per cent. In the Isle of Wight trial, a 21 per cent saving in household use was estimated due to metering. At the same time, a 22 per cent drop in water put into supply was noted, indicating significant reductions in leakage. However, other factors will have influenced customer demands such as; housing density and type, location, weather, occupancy, public awareness of conservation, tariff etc, but these are difficult to quantify.

On the basis of these trials, it is reasonable to conclude that the introduction of universal metering in a similar environment could result in household demand savings of 5 per cent to 10 per cent. Savings in leakage can also be expected.

In the UK, the regulating bodies are promoting the concept of reviewing the possibility of introducing meters on a selective basis to control demand before they are willing to consider applications for new licences, particularly if they are for controversial schemes such as reservoirs or ground water supplies. Some supply companies openly embrace the concept whereas others are less keen suggesting that the effects will be shortlived, representing blips in a steadily rising demand curve, and only postponing the need for new resources by a few years.

Metering will undoubtedly involve significant investment, costing around £200 per household. The tariff will also have a key effect on control of demand; for instance if set too low, it could perhaps increase demand.

There are many other complications such as the question of metering blocks of flats or any other properties sharing supply pipes, and also the cost of maintaining and reading the meters once installed.

The effects on customers' bills and hence their acceptance of metering as a fair system of charging may well be significant in any public debate. Results from the National Metering Trials showed that over 70 per cent of households accepted that it was reasonable to meter water. However, the trials were not wholly representative of the socio-economic distribution nationally. Low rateable value properties do not currently pay a level of charge which is proportionate to their water use, thus the introduction of water meters would mean increasing bills for this group.

In a situation where a significant number of low income households would be paying more for their water, then any public relations exercise would have difficulty in gaining majority acceptance for universal metering.

The effect of metering on demand elsewhere may well be less than experienced in the UK metering trials, particularly if the climate is cooler and wetter than Southern England where most of the trials were located. Less external usage on garden watering etc. will then be evident hence the average per capita demand could be lower.

The economic and environmental benefits of universal metering can outweigh the costs where water resources are scarce, and where development of new sources and infrastructure is more expensive than for an area with adequate resources. For example, Anglian Water have declared their commitment to universal metering, whereas many other UK water companies have not.

Tariffs

In deciding the costs to be recovered from different customers, companies need to allocate costs in a number of ways:

- between water supply, sewerage and drainage
- between metered and unmetered consumers
- between fixed standing charges and measured charges

Most metered tariff structures in operation throughout the world, fall into four groups:

- Fixed charge plus uniform volume charge
- Fixed charge plus decreasing block volume charge
- Fixed charge plus increasing block volume charge
- Fixed charge plus seasonal/peak/volume charge

Uniform volume charges are simple and easily understood.

Decreasing block volume charges is the most common two part tariff and is where the unit volume charge decreases as consumption increases. It is widely used in North America and in some areas in Australia and Western Europe. Declining block tariffs do not encourage water economy and tend to discriminate against small water users.

The use of decreasing blocks has been criticised, particularly in areas where new water sources are expensive to develop. Their use internationally has gradually been reduced during the 1980s.

By contrast, the use of increasing block tariffs (where the volume charge per unit increases as consumption rises) has gained in popularity over the last decade. Examples of such tariffs can now be found in Europe and the USA.

Clearly, as the same supply of low priced water is given irrespective of family size, large families are allowed less per head and these may easily move into higher priced consumption blocks.

Increasing block tariffs are used to curb demands and are designed to ensure that customers demanding large volumes of water realise the high costs involved in developing new supply sources. Japanese water undertakings, have adopted this rationale in the design of their tariffs.

The seasonal tariff is today gaining in popularity, particularly in the United States. The basic idea is that, since peak summer consumption costs more to meet than winter consumption, then the price should reflect these excess costs.

One scheme is to combine block and seasonal tariffs. Winter use provides the 'block' and only excess consumption in the summer over winter volumes is charged at the higher rate.

Although it is superficially attractive, this tariff does not seem an improvement over the simple season tariff. A cubic metre saved in the summer is of equal value whether saved by someone with high winter consumption or low winter consumption. It discriminates against those who, for whatever reason, use little water in the winter. Indeed, there can be perverse incentives to consume a little more in the winter. This tariff structure is probably of less advantage to large families than appears on casual inspection.

It is argued that customers with meters should have the genuine opportunity to influence their bills, hence fixed charges should be low. But the disadvantage to the companies with this approach is that revenue is more uncertain because seasonal demands and economic recessions become considerable risks.

Some suppression of demand is therefore expected if the sliding scale of tariffs is pitched correctly.

Comprehensive debate (Ref 3) is ongoing in the UK water industry on the subject 'Paying for Growth'. Reference can be made to the related documentation for details and viewpoints including those on charges for measured and unmeasured consumption.

Whatever tariff structure is adopted, it needs to achieve:
- fairness and equity
- sensible incentives to customers and to companies
- simplicity and comprehensibility

Water using appliances

In the UK, water used in various household activities is typically (litres/head/day):

- toilets   35
- clothes washing   20
- dishwasher   11
- baths and showers   36
- outside   9
- cooking, cleaning, drinking etc   35
                                              146

Toilets and washing therefore currently account for about 70 per cent of average household daily consumption. In any assessment of future demands, changes in the rate of use of water using appliances and the volume per use must be taken into account.

In the UK, it is generally considered that toilet usage will remain static, but that the volume per use will reduce gradually as new houses are built and existing systems are replaced with the lower capacity cisterns now required by the 1991 Water Regulations. An average reduction after 10 years of only about 1 l/hd/day is expected.

Bath and shower use is often forecast to increase due to increased emphasis in personal cleanliness and the likely increase in the number of power showers. Usage in showers and baths is affected primarily by personal habits and preferences. While household supplies are unmeasured, no major constraint on the forecast growth in this component of usage is foreseen.

In the UK, the volume per use of both automatic washing machines and dishwashers is expected to reduce by around 25 per cent. However, increased ownership and usage are expected to offset these reductions.

The size of washing machines and dishwashers is determined by the manufacturers who are more constrained by reducing operating cost, i.e. power, rather than conserving water. European manufacturers may be more affected by 'green' policies and customer support for

conservation. If all machines are limited to approximately 85 litres/wash, it may be possible to reduce demand by about 3 l/hd/day over their 10 year life. Until water consumption becomes a significant element in the choice of machine, there is little incentive on manufacturers or consumers to achieve these savings.

Apart from the above, it is generally concluded that there are no real opportunities for further demand management of water using domestic appliances.

Consumer education

Publicity campaigns give the opportunity for public education in the work of the water companies and to make people more aware of the effects of wastage and extravagant usage. In a drought situation they should be used as a preliminary stage prior to the introduction of restrictions.

Advice can be given on how much water is needed on a garden, and perhaps the public can be encouraged more to use native species of plants and water butts to reduce water taken from the distribution system.

In discussions with system managers, there is often general agreement that publicity campaigns are an important part of public relations, but that they make little difference to the level of demand.

Commercial and industrial customers in both the private and public sectors are usually very aware of the cost of water. In one area of UK, the dual effect of increasing water charges and the introduction of trade effluent charges have resulted in significant effort by major customers to reduce leakage and waste and to increase recycling of process water. The effect of this is expected to reduce industrial and public sector demand over the next 10 years, and thereafter to reduce the rate of growth as new industries have more water efficient processes.

In commercial premises, retrofit devices are now being used to prevent wastage, particularly in toilets.

Leakage and waste prevention

Considerable effort is now being expended in an attempt to understand what happens to the water after it leaves the treatment works.

Terms such as 'unaccounted for water' are being phased out as new terminology such as 'water delivered' and 'distribution losses' are adopted. Water delivered is, what it implies, water delivered by the water company to the consumer. It has many sub categories — such as domestic unmeasured, domestic measured, commercial non measured etc.

'Distribution losses' includes leakage upstream of the 'point of delivery' to the customer and is the area where the water company can, by a structured and organised approach, implement system and management policies to save water.

Leakage and waste are significant components of distribution input. Measures which can reduce losses will be of considerable benefit since leakage control has an influence on customer charges, financing of future growth and allocation of water resources. Those factors which can influence leakage are summarized as:

- o pressure management
- o refurbishment of mains
- o replacement of corroding service pipes
- o refurbishment of internal plumbing systems
- o leakage control programme

In areas of high pressure, reduction in pressure can produce significant savings in leakage and reduce wastage in the house. Reductions in pressure from 70m to 40m can halve leakage levels. Pressure reduction should be a high priority where it is possible.

Refurbishment of mains and replacement of service pipes is more likely to be driven by the requirement of water quality. However, significant proportions (in the region of 75 per cent) of distribution leakage is thought to come from service pipes. Such rehabilitation measures will therefore produce reductions in leakage.

Leaks on customers' properties are not the responsibility of the water company. When leaks are detected, notices can be issued instructing the customer to organize repairs. No significant improvement to this system is usually envisaged unless the leak is downstream of the water meter and the customer realizes he is paying for wastage.

Improvements in the technology available to monitor and detect leaks will reduce the level of leakage. Continual night flow monitoring is becoming the standard practice. By setting up district meter areas, and then monitoring the flow into the area each night (either by telemetry or by a locally placed logger downloaded at regular intervals), any change in net night flow can be quickly identified and further investigative work carried out to pinpoint the leak. Recent research work has high-lighted that although major bursts lose high rates of water for short periods, more water is lost through smaller leaks running over much longer periods. Hence, it is important to reduce the period over which these leaks are allowed to run. The Industry textbook on leakage is the STC Report 26 (Ref 2). This book, now 15 years old, introduced many concepts, but is becoming obsolete due to technical advances in the industry, and there is currently a National Leakage Initiative (NLI) which is updating Report 26.

The NLI is producing some interesting results, including the need to quote leakage in terms of $m^3/day/km$ rather than m^3/day, and the impact of distribution pressure on leakage. This very interesting work is ongoing and will be fully reported in the near future.

Summarizing, demand management, in particular leakage control, has come to the forefront of the water industry's thinking. It is now a high profile subject and more information is becoming available. What remains undisputed however, is that a successful demand management and leakage control programme will reduce operating costs, enable better use of resources and defer capital expenditure, all to the benefit of the investor.

References and acknowledgements :

1. The National Metering Trials Working Group: WSA, WCA, OFWAT, WRc, DoE, 1993

2. UK Water Industry National Leakage Control Initiative

3. Paying for Growth : Consultation Paper, Office of Water Services, February 1993 and subsequent support documents.

Acknowledgement is given to Mr CJA Binnie : Managing Director, Mr TEA Askew : Divisional Manager (Water) and Mr JA Downer (Engineer), all of WS Atkins Consultants Ltd, for their support in the preparation of this paper.

Project management of borehole programme

David Van Der Westhuijzen and Kalinga Pelpola, EVN projects management, South Africa

BOPHUTHATSWANA GAINED INDEPENDENCE from South Africa in 1977 and became one of the fastest growing developing countries in the region.

Prior to the establishment of the Department of Water Affairs (DWA) in 1987, more emphasis had been placed on urban water supply, whilst an ad-hoc approach was followed for rural supply.

In 1987, it was estimated that nearly 1.5 million people in the rural villages of Bophuthatswana had no reasonable access to potable water. The Department of Water Affairs adopted a strategy which relied on requests from the rural community for boreholes, rather than attempting to supply water to all areas on a pro-active basis. This system of properly verified and prioritized requests proved to indicate those areas with the greatest need for water.

Between 1987 and 1989, the Department of Water Affairs attempted to satisfy requests for water with departmental plant and personnel. Lack of adequately trained staff and frequent breakdowns however soon rendered the operation ineffective. Work output and success rates were low and targets could not be met.

This was development for the people, on behalf of the people... anything but development by the people.

Any rural water supply scheme making use of groundwater resources, involves two key role players:

- **The Project Sponsor** — This may be Government, through one of its departments, or it may be a non-governmental organization.
- **The Community** — Those consumers in need of water.

The traditional rural water supply programmes were characterized by the initiative being vested in the Project Sponsor. The then Republic of Bophuthatswana, was no exception.

Most rural water supply projects involved consultants and contractors. All of these wedged in between the project sponsor and the community. This results in management gaps — limiting the essential lines of communication between the various levels of management. (Refer Figure 1)

To add to the dilemma, there also exists functional gaps between the various working groups in the team. For instance between the geohydrological and engineering consultants, or between the drilling, testing and equipping contractors. If we superimpose the management gaps on top of the functional gaps, we find that traditional rural water supply teams are made up of small operational islands. (Refer Figure 1)

The Bophuthatswana Rural Water Supply Programme is operational in all five regions of Bophuthatswana with separate contracts for drilling, testing and equipping of boreholes in each region. Allowing for the departmental staff and four consulting firms, approximately 200 people are directly involved in the programme on a daily basis.

A further dimension to the problem is the geographical location of the various regions of Bophuthatswana.

- Five regions, 14 districts, although only 44 500km², it is spread over an area ranging between 5 and 600 km from the capital Mmabatho, housing a 1993 population of 2,3 million people.

The result of this traditional approach in Bophuthatswana were inevitable:
- hundreds (or even thousands) of boreholes;
- no database or borehole numbers;
- no testing records;
- no follow-up on equipping and maintenance.

We can add to this:
- no involvement of the community;
- no sense of ownership.

The Bophuthatswana Rural Water Supply Programme was launched by the Department of Water Affairs in 1989.

The long-term objectives of the programme were defined as follows:

- To provide every Bophuthatswana resident with safe, affordable and sustainable drinking water at a rate of 25 litre per capita per day, within 500m from his dwelling, before the year 2000.
- When the programme objective of Water for all by the year 2000 was defined in 1989, it was estimated that water had to be supplied to an additional 85 000 people per year. The actual progress has exceeded this target each year since 1989.

Subsequently, these objectives have been amended to include water for agricultural use as well as the installation of yard connections in villages where a sustainable source of water is available.

The primary objective of this programme was therefore to provide every resident who has already expressed a need, with a safe, sustainable and affordable supply of water. Water was supplied to villages, schools, clinics, hospitals as well as for communal agricultural use.

The principle difference between the traditional approach and the Bophuthatswana Rural Water Supply Programme was this:

The initiative was vested in the community — they become the core of the programme.

The new strategy broke away from departmental attempts at siting, drilling, testing and equipping of boreholes and included the appointment of project co-ordinator, consulting engineers and geohydrological consultants.

The project co-ordinator was responsible for co-ordinating and integrating activities across multiple, functional levels. Project management principles of plan, organise, lead and control, are therefore the ingredients that cement these operational islands into a continent of co-ordinated effort.

Based on the historical expenditure on contractors under the programme, the unit cost per person served, is estimated to be in the order of only US$ 15.20. This is in sharp contract to the annual cost of US$ 41.44 per person normally paid for water when purchased from private vendors.

Within the limits of time, cost and performance, all these efforts are directed towards the common goal:

Water for the people, by the people.

For a sustainable supply of water a borehole must be introduced into the community with a maximum of local involvement.

The programme was request driven. Any representative of the community was allowed to submit a request for communal water.

Upon receipt of a request, members of the management team verified such a request to establish whether it is a bona fide request or not. This included a visit and a survey of existing water sources to determine whether:

- there was no existing or alternative sources of water within 500m from the desired location of the source;
- the need has not been satisfied in the meantime by somebody else;
- the request could not stand over until a regional scheme is supplied, i.e. urgency or priority of request.

Where more requests were received than could be accommodated in the budget, prioritization was done to determine which must be satisfied first.

Prioritization was normally done on a village basis, after which priorities between villages were assigned.

The activities of siting, drilling, testing, design and construction for each verified request are scheduled according to the priorities within the contents of the district budget.

The consultants and contractors proceed with the execution of the scheduled activities:

- Siting of boreholes was executed by professional geohydrologists to ensure the highest probability of

successful drilling. The latest scientific methods were applied in a cost-effective manner.

- Boreholes were drilled by approved drilling contractors selected through public tender. Boreholes were drilled, lined and developed under the supervision of a geohydrologist.
- Boreholes were tested to establish the sustainable yield as well as the water quality. Testing contractors were selected through public tender. Boreholes were tested under the supervision of a geohydrologist who issued a management recommendation for each borehole.
- A civil engineer designed each piece of equipment in accordance with the management recommendation for the specific borehole. The equipment were handpumps, windmills, tanks, motorized pumps and minor pipelines as may be appropriate in each case.
- Construction activities were labour-based as far as economically viable. Equipping contractors were selected through public tender. On completion, all equipment will be handed over to the community for operation and maintenance.

Communication is a cornerstone of a successful programme. The two most important types of formal communication are meetings and standard forms.

Formal meetings are held at four levels:

- Department of Water Affairs Technical Appraisal — policy decisions, approval of reticulated schemes.
- Rural Water Supply Programme Management Team — policy and strategy.
- Project Co-ordination — overall programme co-ordination and progress.
- Regional Co-ordination — programme co-ordination at regional level, contractor site meetings, information to and feedback from regional community representatives and other functional departments.

Minutes of these meetings, as well as all supporting documents to progress reports constitute the written formal communications in this category.

The use of standard forms is a simple yet powerful means of communication. By carefully designing these forms, a wealth of information can be communicated. This method has the following advantages:

- simple (usually single A4 size page);
- easy to understand and refer to;
- can act as a checklist;
- can act as information interchange as well as work authorization;
- backup to computerized database.

No fewer than 19 standard forms were specifically designed for use on RSWP, all properly identified and referenced by number. Whilst these forms are standardized, they are constantly under review and scrutiny for the introduction of improvements and multiutility. In

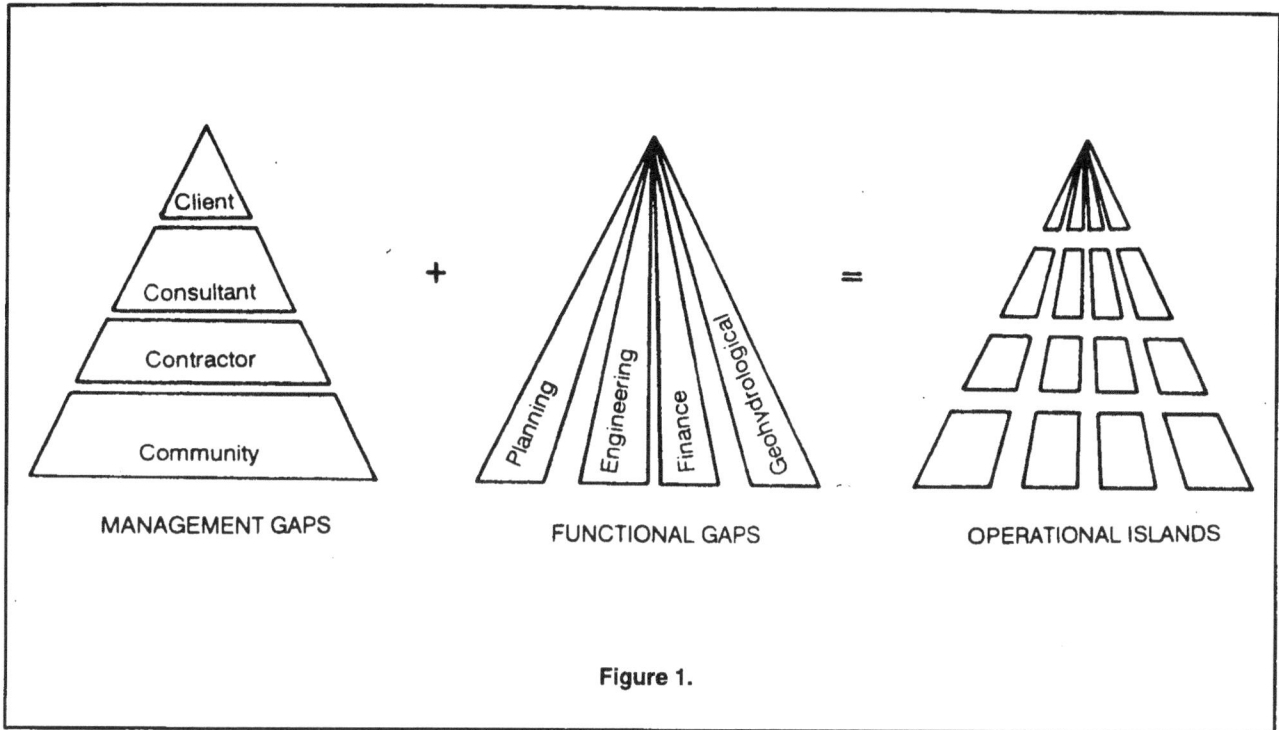

Figure 1.

many cases, standard forms have replaced lengthy written reports.

The mere accounting and reporting of the cost after the action has occurred cannot in itself control any costs. Historic data can however be used to predict future costs, and this formed the basis of the Rural Water Supply Programme cost control system.

A streamlined administration system has been specifically developed to handle an average turnover of approximately R5 million per month.

Actual and budgeted unit costs were used to forecast future costs, which taken within the constraints of the annual budget, allows a certain amount of work to be scheduled in the format of monthly Gantt charts.

Each one of the 4500 requests received, resembled a small but unique project on its own. Each request involves site visits, borehole siting, drilling, testing, designs, constructions and even completion certificates. A total of six activities are managed per request resulting in a total of 27 000 activities on a bar-chart (each one with its unique dependencies and resources).

This allows the civil engineer to prepare work authorization sheets and technical designs timeously, so that contractors may order materials ahead of time and thus avoid possible delays. Costs estimates of the work to be performed are then fine-tuned and taken as committed funds, against which actual costs can then be measured, as well as the effect of committed funds on the overall budget.

Gantt charts are summaries of the logic network and trace the progress of requests from request stage to handover. The Gantt chart is a powerful scheduling tool, and forms the basis of monthly regional co-ordination meetings. Whilst the district budget dictates how many boreholes may be drilled, tested or equipped in the following month, the Gantt chart defines precisely which boreholes are to be actioned.

The growth of Rural Water Supply Programme can largely be attributed to an increased implementation capacity through more effective and efficient management of the programme, but it must be acknowledged that the availability of funds removed constraints which otherwise may have impeded progress. In the period August 1989 to the present, the management system of Rural Water Supply Programme has been going through a constant process of evaluation and fine-tuning, necessitated by the massive amounts of information generated on a programme of this nature. Effective management hinges on the capturing, processing, assimilation and interpretation of up-to-date, complete and accurate information, which of course leads ultimately to intelligent and timeous decision making.

Upon completion of the work related to a request, all pertinent information is recorded in a locality particular computerized database. This database interacts with CAD and GIS to facilitate mapping and plotting.

Apart from borehole information generated as part of Bophuthatswana Rural Water Supply Programme, information concerning existing public boreholes is constantly being collected in a national borehole census. At present there are in excess of 7000 boreholes recorded on the database.

Since 1989, the following progress was made:
- Requests satisfied   3775
- Boreholes drilled   2400
- People provided with basic water   750 000

The success of the programme is seated in the strong partnership between the public service, the professional advisors and the communities. All efforts are directed towards satisfying the real needs defined by the people themselves. According to a survey conducted by the RSA Department of Water Affairs, the rural communities of Bophuthatswana are the only areas (apart from Qua-Qua) where availability of water is higher than 15 litres per person per day.

Through the establishment and training of village water committees as well as the creation of employment during construction, the monies are spent in a manner that results in the greatest benefits to these communities.

References

[1] Kerzner Harold, *Project Management: A systems approach to Planning, Scheduling and Controlling*, Van Nostrand Reinhold, 1989.

[2] Moder, Joseph J, Cecil R. Phillips, Edward W. Davis, *Project Management with CPM, Pert and Precedence diagramming*, Van Nostrand Reinhold Company, 1983.

[3] Department of Water Affairs, Bophuthatswana Rural Water Supply Programme, Draft Manual, October 1992.

[4] *African National Congress, The Reconstruction and Development Programme, (A policy framework)*, Umanyano Publications, 1994.

[5] Field experience of the Bophuthatswana Rural Water Supply Project Team.

Quality of drinking water

Eng. (Mrs.) C. Wethasinghe and K.M. Manickavasagar, NBRO, Colombo, Sri Lanka

IN SRI LANKA the entire drinking water supply system is managed by the National Water Supply and Drainage Board (NWS&DB). Hence it is necessary that an independent organization carry out a surveillance programme for pipe-borne drinking water quality. As an independent agency the National Building Research Organization (NBRO) in Colombo, Sri Lanka, commenced the on-going Water Quality Surveillance Programme for pipe-borne drinking water, supplied to the Greater Colombo area in May 1988, at the request of the NWS&DB. The full scale operations for the entire Greater Colombo area were met by September 1988. The objectives of this programme are;

- to survey the quality of water supplied through standposts, taps in households, taps in public places, overhead tanks/sumps and taps in housing schemes
- to provide information to the NWS&DB on the quality of water for remedial action
- to improve the reliability and the quality of supply of water and to create a general public health awareness of clean and safe water.

Surveillance criteria

This programme is carried out in accordance with the criteria stipulated by World Health Organization (WHO). WHO has defined criteria in terms of sample collection based on population (Table 1). These criteria were adopted in deciding the number of samples to be collected in each area.

The Water Quality Surveillance Programme covers the following areas in the Greater Colombo region; Colombo City, Dehiwela-Mount Lavinia-Ratmalana, Moratuwa, Panadura, Sri Jayawardanapura-Kotte-Kolonnawa and supply and service reservoirs at Kolonnawa, Mulleriyawa,

Labugama, Kalatuwawa, Churchill, Elihouse, Maligakanda, Dehiwela and Panadura.

The programme includes collection of water samples from various points in the distribution system; standposts, households and public utilities. Samples are also collected from the storage system; overhead tanks/sumps and housing schemes. Samples from reservoirs are also analysed.

The sample estimations in the distribution system for each area were determined according to the above WHO criteria and are given in Table 2.

Biological examination offers the most sensitive test for the detection of recent and potentially dangerous fecal pollution in drinking water supply system. It is essential that water is examined regularly and frequently as contamination may be intermittent and may not be detected by the examination of a single sample. Hence it is important that drinking water be examined by a simple test or series of tests. According to WHO guidelines priority must always be given to ensure that routine bacteriological examination is maintained whenever manpower and facilities are limited.

The standards set for drinking water quality in order to minimise public health risks have been based on the use of total coliforms, fecal coliforms, and faecal streptococci as indicators. Since it is very complex and often costly to perform virological analysis, the bacterial indicators continue to be the most reliable gauge of hygienic quality of water.

The membrane filtration technique is used for identification of coliforms.

All the samples are tested for faecal coliforms and residual chlorine. Though pH was determined for all the samples collected during the initial stage of the programme, at present it is carried out for only one sample

Table 1: World Health Organization criteria for sample estimation

Population served	Minimum number of samples to be taken from the distribution system each month
<5000	1 sample/month
5000-100,000	1 sample/5000 Population
>100,000	1 sample/10 000 Population

Table 2: Sample estimation

Area	Number of Samples Collected/mopnth
Colombo City	70
Dehiwela-Mount Lavinia-Ratmalana	20
Moratuwa	15
Panadura	10
Sri Jayawardanapura-Kotte-Kolonnawa	20

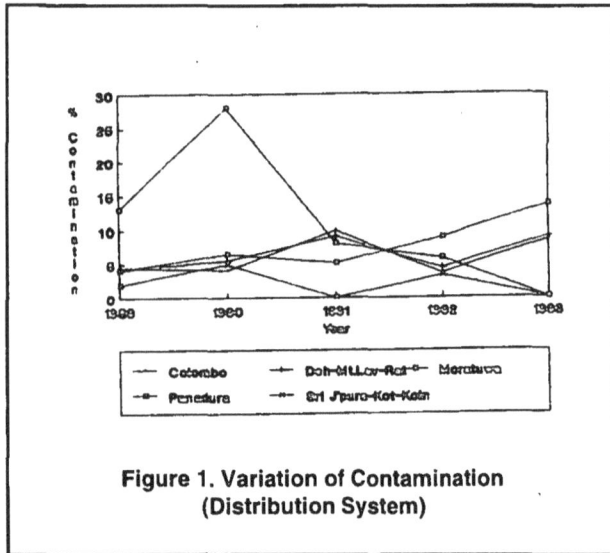

Figure 1. Variation of Contamination
(Distribution System)

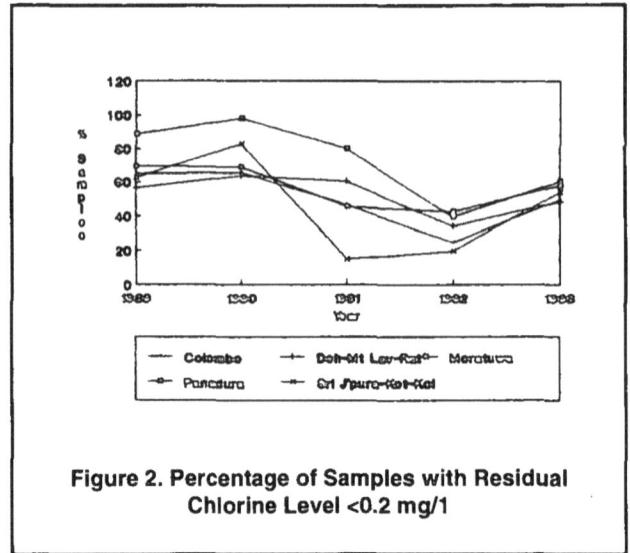

Figure 2. Percentage of Samples with Residual
Chlorine Level <0.2 mg/1

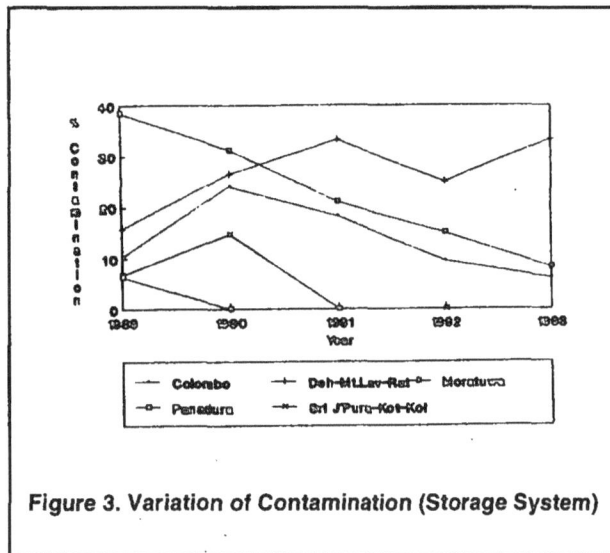

Figure 3. Variation of Contamination (Storage System)

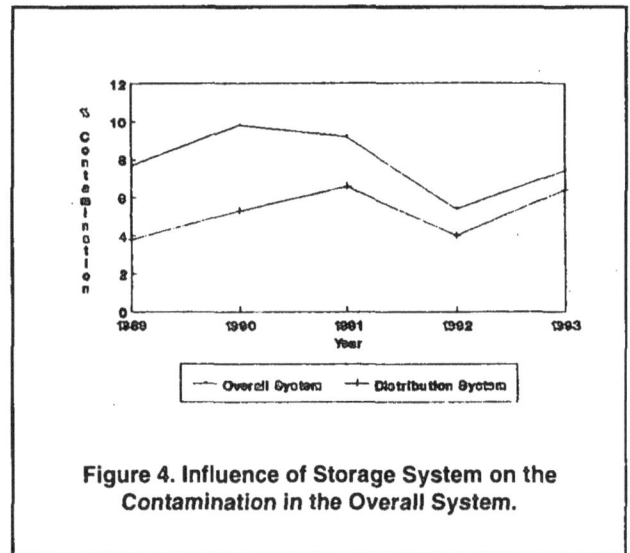

Figure 4. Influence of Storage System on the
Contamination in the Overall System.

collected from each zone. The samples from supply and service reservoirs are tested for faecal and total coliforms, residual chlorine, pH, iron and alkalinity.

Method of operation

Routine sampling is carried out according to a schedule prepared at the beginning of each month. The results are reported in specially formed data sheets.

On detection of faecal contamination in a sample (suspect contamination), re-sampling of the same point is carried out within the next 24 hours. Further detection of contamination is considered confirmed faecal presence., and all the confirmed faecal presence are immediately reported to the NWS&DB for corrective action. However if more than 75 per cent of the samples collected from any zone showed suspect contamination, re-sampling is not carried out and results are immediately reported to NWS&DB as confirmed faecal presence.

At the end of each month, the NWS&DB is provided with a monthly report including the test results and area summaries.

All data are currently being processed on Lotus 123.

Evaluation of results

The data collected during the period January 1989 to December 1993 were analysed and evaluated in this paper.

Distribution system
The distribution system includes the samples collected from standposts, households and public utilities.

Standposts
Outlets of water from the distribution system by means of a standpost.

Household
Water supplied to the households directly from the mains.

Public utilities

Water used directly from the distribution system in public places such as restaurants, hotels, hospitals, etc.

A total of 6203 samples have been tested from the distribution system during 1989 to 1993 and 384 contaminations from all sources in the system were detected. The area-wise distribution of this contamination is presented in Figure 1. The highest percentage contamination of 28 was recorded in Panadura in 1990.

The analysis of data indicates a low residual chlorine level in Pandura area where the highest contamination was also recorded (Figure 2). Also observed is a reduction of percentage of samples with chlorine level less than 0.2 mg/l in all the areas.

Storage system

Out of 1132 samples tested from the storage system during the period 1989 to 1993, 200 contaminations have been detected. The variation of this contamination is given in Figure 3. According to Figure 4 there is a significant contribution from the storage system to the overall contamination. However the graph indicates a reduction of contamination due to storage system at present.

Conclusion

The results indicate a direct correlation between the level of residual chlorine and the degree of contamination.

A positive downward trend in the contamination of storage system is currently observed consequent to highlighting the results at health committees in the past.

The feedback from MBRO to NWS&DB on the quality of water and their remedial action has improved the level of residual chlorine in the distribution system.

Acknowledgement

The authors wish to acknowledge the staff of Environmental Division of National Building Research Organization for their contribution in this programme.

References

WHO Guidelines for Drinking Water Quality, Vol I, II and III.

Are handpumps really affordable?

Michael Wood, CARE, Ethiopia

DURING THE INTERNATIONAL Drinking Water Supply and Sanitation Decade of the 1980s many thousands of handpumps were installed in developing countries as part of the United Nations-led drive to provide safe drinking water and adequate sanitation for all by 1990.

Since then, thousands more handpumps of many different types have been put in by donors and governments.

Handpumps have been given a high profile in the quest to provide potable water to the world's burgeoning rural population by leading players in development like the World Bank, UNICEF and a plethora of international non-government organizations. Handpumps were vigorously promoted as being the best option by which communities could enjoy a safe and reliable water supply, based on the following set of assumptions:

That handpumps were:

o Low cost
o Affordable
o Easy to maintain
o An appropriate technology
o Readily available
o Easy to install
o User friendly
o Efficient

It has been generally accepted that handpumps render a shallow well or borehole safe against surface contamination based on the belief that the water will be contaminated to an unacceptable degree (having an E-Coli count of more than 10) if alternative extraction methods are used e.g. rope and bucket.

This paper will point out that handpumps have not lived up to earlier expectations, particularly in the area of affordability, and that donors and recipient governments would be well advised to consider other less politically correct options, under certain circumstances.

A brief history of handpumps

The first generation of handpumps included such stalwarts as the British made Godwin, which were installed in the 1930s, 40s and 50s. They used super heavy duty materials like cast iron and hardened steel in the belief that one had, in the colonial era, to make pumps virtually indestructible so as to withstand constant use and abuse by people in the Third World who could not be expected to maintain, let alone repair, such advanced pieces of technology.

To their credit, many such pumps continued to pump water for many years beyond their original life expectancy, but many also broke down and stayed that way for months or years because government mechanics did not come to repair them for a variety of well-documented reasons.

During the 1960s and 70s a second generation of handpumps emerged, of which the India Mark II is the most notable example. With over five million installed worldwide, this is undoubtably the world's most widely used handpump. At the time of its development in India in the early 1970s, the Mark II was heralded as being the answer to the myriad problems of rural water supply.

However, this pump relied on a three-tier maintenance system. Although such a system was developed in India, in 1986 it was reported (World Water Conference, Nairobi) that over one million India Mark II pumps were broken down on the sub-continent.

The pump was, however, considered an affordable option at least in the Indian context where intense competition in the burgeoning manufacturing sector kept costs down.

It is still probably the most cost-effective handpump for depths up to 45 metres, even in Africa where high freight costs have always made imported pumps more expensive than they are in India.

But in Africa, the Mark II has not been a sustainable solution to rural water supply problems, mainly because the Indian-style tiered maintenance system frequently failed or simply was not there in the first place.

Development of the Afridev

During the 1970s, the World Bank/UNDP pioneered the concept of a simple handpump which could be maintained at the village level in Africa. The Bank financed the development of what became known as the Afridev, based on the belief that handpumps must be made and maintained locally, by local people. The Afridev design featured state-of-the-art lightweight, non-corrosive, easy-to-assemble materials developed in co-operation with the Swiss multinational company, Dupont. Ironically, manufacture of the Afridev has never really taken off in Africa. One of the reasons being the extremely high price of the mould needed to produce the nylon bearing-bushes and the footvalve/plunger. Also, high import tariffs on raw materials make the manufacture of Afridevs in African countries expensive.

The Afridev is, however, being made in large numbers at competitive prices in India and Pakistan and is being sold for installation in African countries cheaper than African-made Afridevs !

One of the main reasons is that in most African countries the small scale industrial base is not nearly as developed as is the case in India or Pakistan. The price of a European-made Afridev landed in an African country right now is about US$900.00; about double what an Indian-made Afridev costs!

The handpump option

The donor community throughout the Water Decade, did much to persuade governments of developing countries that handpumps per se offered the best option in making safe water available to burgeoning rural populations.

The advent of the Village Level Operation and Maintenance handpumps in the late 1970s to early 1980s did much to further the handpump option, particularly in Africa with the Afridev leading the way toward the goal of affordable village-based maintenance.

Are handpumps sustainable?

Sustainability may be defined as an intervention which is capable of being supported and maintained by a community or individual over an extended period of time with an absolute minimum of outside assistance.

VLOM handpumps were developed and installed in remote rural areas because it was assumed that the users themselves would be able to maintain them. In many cases in Africa this has proved impractical due to a number of technical problems with the Afridev pump concerning:

- The PVC rising main
- The method of joining pump rods
- The nitrile rubber seal and O ring
- Fishing tools
- The supply of spare parts kits

Rising main

The Afridev blueprint specifies a 63mm O.D. PVC riser pipe having bell joints glued together. Originally it was thought that it would be unnecessary to remove these pipes once installed in the well. This is a big selling point. However, experience in Malawi, Ghana and Ethiopia has revealed that in some types of Afridev, the rod connector wears a hole in the riser pipes, necessitating their removal by sawing and re-gluing using PVC sockets. This operation is beyond the means of handpump caretakers. Also, PVC risers installed in wide-diameter wells tend to flex during pumping causing joint fatigue leading to cracking of the PVC pipe. Little thought has been given as to how to secure PVC pipes in the well.

Pump rod joining

Some manufacturers use a plastic clip-on device for joining the rods. These can and do come off after a few months use, necessitating the use of a **fishing tool** to extricate the fallen rods. The type of fishing tool supplied by the manufacturer is not able to do this job, so a special tool has to be fabricated. This too is beyond the means of handpump caretakers in the village.

Plunger seal and O ring

Experience has shown that the nitrile rubber plunger seal and footvalve O ring actually absorb water over time, and expand. This makes their removal difficult, especially in the case of removing the footvalve.

Supply of spare-parts kits

It is recommended by Afridev manufacturers that the plunger seal, bearing bushes and footvalve bobbins and O ring be replaced annually as a preventive maintenance strategy. However, the issue of how the spare-parts kits are going to be supplied to the village caretaker has not been fully addressed. Difficulties arise when donors try to supply spare parts at the village level. Who is to look after them? Is the village expected to pay? Who is going to collect and keep the money? Are parts to be given freely or should a nominal charge be levied ?

Handpump caretakers

Many thousands of handpump caretakers, many of them women, have been trained to maintain handpumps like the Afridev. But this pump still has its problems. Can caretakers and their assistants fully repair this type of handpump? Experience to date suggests that they cannot. Most water supply projects have convenient showpiece communities not far from project headquarters where groups of highly trained women impress visitors by whipping out the rods and changing the plunger seal in textbook fashion.

What is less well known, but just as common, is that VLOM handpumps have failed in remote rural areas because problems have arisen beyond the means of the trained caretakers to repair.

Beautiful wells have been rendered useless and people have been forced back to traditional, unprotected sources because the VLOM handpumps have broken down, typically with rising main problems.

In Africa, the India Mark II does not enjoy an impressive record of sustainability mainly because there are not the village level mechanics available that are commonly found in rural areas of India, where the popularity of the ubiquitous bicycle has encouraged a culture of village bicycle repair shops whose mechanics are ideally suited to repair handpumps, a technology on a par with that of bicycles. As the bicycle makes inroads into the African countryside we can expect an upsurge in the repair business which will augur well for the continued sustainability of handpumps.

Affordability of handpumps

When we talk of affordability we must ask, affordable to whom? Who is paying? It has been said, for example, that the Afridev is an affordable handpump for Ethiopia. (Second National Handpumps Workshop, Addis Ababa, UNICEF, Jan.92). But is it?

In 1992 a bilateral aid project in Ethiopia imported 165 Afridevs from India costing US$660 each including airfreight. If the cost of clearing, transport to the site and installation costs are included, the installed cost comes to around $700. Each handpump serves about 55 households. The World Bank states that the average per capita income in Ethiopia is $120. Therefore, if the users were paying, each family head would have to pay $12.72 or 11 per cent of their annual income. This is more than double the five per cent guideline that the Bank has said is the maximum that families should have to pay for safe water. Clearly, in this scenario handpumps were not an affordable option.

The technical shortcomings mentioned above and the expense involved call into question whether handpumps, such as the Afridev, are the most appropriate and sustainable solution to potable water supply problems in rural areas. In isolated rural communities in the Tigrayan mountains of Ethiopia, or across the savannah lands of the Sahel, where outside technical assistance may be weeks or months in arriving, communities have been left without a safe water supply because their so-called VLOM pump failed and they couldn't fix it.

Alternatives

Having shown that handpumps may not be affordable or technically sustainable, we have to ask, What is the alternative?

One answer is Back to Basics!

Back, in fact, to the age-old rope and bucket system. But that is not the whole solution if contamination is to be kept to a minimum.

Coupled with this simple approach, must come improved well-head design featuring a large, well-drained concrete apron, a protective parapet and a simple windlass to which the rope is attached. Having a dedicated bucket, be it half an inner tube or a proper bucket, will further reduce the risk of contamination.

Hygiene education

An integral part of a rope and bucket system must also incorporate a hard-hitting hygiene education program which focusses on women, the main users and managers of household water, and children, who are the most receptive to behavioural change.

Hygiene messages should be simple, to the point and unambiguous. The user community must be targeted with a well-thought out, ongoing education program, NOT just a blitz-like campaign that is here today gone tomorrow.

VLOM handpumps cost from $US400 to 800 each. The money saved by NOT installing a handpump could be used:

- To finance improved well-head works
- To conduct ongoing and effective hygiene education programs
- To build more wells in other communities thus making potentially safe water more available to more people.

Conclusions

Some people will undoubtably think that the rope and bucket system is taking a step backwards; that it is too primitive; that rural communities deserve something better.

In some cases handpumps are indeed viable and sustainable, even affordable. But in many isolated rural communities in the emerging countries of the South, many millions of people still live and die of preventable diseases associated with unsafe water supply and inadequate sanitation facilities.

In order to increase coverage; to scale up to the levels required, we should not put all our faith in handpumps, but rather concentrate on more sustainable and more affordable solutions to the problems of rural water supply, so as to transform the goal of Health for All by the year 2000 into an achievable reality.

SECTION 7

WATER TREATMENT

An appropriate iron removal technology

M.A.M.S.L. Attanayake and J.P. Padmasiri, Kandy District, Sri Lanka

EXCESSIVE IRON IN groundwater became a major setback for the continuation of the hand pump well programme of the FINNIDA assisted water project in Kandy during mid eighties. As a result of very high user sensitivity a considerable rejection level of handpumps was observed. Since then few types of iron removal plants (IRPS) employing different technologies were installed and tried out in the field between 1985-1987. During the testing FINNIDA Circular IRP type employing the treatment philosophy of biological iron removal showed promising results to bring down the iron content to an acceptable level to the user group. The main limitation in this model was seen as failure of maintenance, in the absence of a centralized repairing team. Based on the experience gained in testing, FINNIDA Square IRP was developed. Field testing was continued subsequently since 1988.

The main design consideration in this development had been community adaptability of the cleaning process with special features such as simplicity in construction, operation and maintenance. Also utilization of material and skills at village level were optimized. Field trials carried out for longer years has proved FINNIDA Square Type IRP as an appropriate technological option for iron removal at community level.

Background

In Kandy District a total of 2500 handpump wells had been completed by Kandy District Water Supply and Sanitation Project under FINNIDA assistance. High rejection rates of the handpump well programme by the recipient community was observed in mid eighties. This was due to high user sensitivity as a result of excessive iron content in water. There were two categories of causes for the problem namely:

(a) Occurrence of iron due to corrosion of below ground component of the handpump well in aggressive ground water.
(b) Occurrence of iron in ground water.

The iron problem due to corrosion was solved by changing the below ground components of the handpumps with new non-corrosive parts. And the problem due to occurrence of iron in groundwater had to be dealt with an appropriate treatment technology at hand pump level. Few types of iron removal plants were installed and tested in order to solve this problem. Some limitations were observed in the application of these technological

options. Therefore R&D work continued in order to establish a more appropriate iron removal technology manageable at community level for the future sustenance of the handpump well programme.

Experience on previous field trials (1985-1987)

Following types of iron removal plants had been installed at different locations and performance was monitored accordingly.

(a) FINNIDA Dual Unit Type
(b) FINNIDA Circular Type
(c) UNICEF Type

Different type of filter media such as charcoal, cement coated polystyrene balls leca particles were utilized in the upward filter for both Dual unit type and the circular type employing biological iron removal as the main operation philosophy. Out of these iron removal plants FINNIDA circular type has shown promising results bringing down the iron levels to the required standards.

However the main limitation observed in this model was the difficulty of frequent cleaning as the cover slab was too heavy and difficult to handle safely. Therefore many installations were dependent on the centralized maintenance unit, making maintenance expensive and cumbersome. The cement coated polystyrene balls used as a substitute for the imported leca pebbles used in the upward filter did not appear either to be durable in the long run or replaceable practically at the village level.

Development of a more appropriate IRP

Finnida square type IRP had been designed as a treatment unit meeting VLOM requirements based on the above experience in the iron removal. The water from the handpump wells was allowed to enter 700mm long 75mm diameter PVC pipes where the aeration will take place due to turbulence caused as a result of vanes provided.

The aeration process had the added advantage where the water column in the aeration pipe had a fluctuation of 300mm height over every discharge making aeration more effective.

Another PVC pipe of 50mm dia. with circular ports staggered at 60mm spacing was connected to the 75mm pipe to distribute the water evenly in the filter. Granite chips 10-25mm size were utilized as the filter media in the upward filter known as the Hopper chamber.

The main feature in this treatment unit is allowing a growth of thriving band of iron bacteria in the upper layers of granite chips. The pH condition at the Hopper chamber becomes favourable for both microbiological oxidation as well as conversion of ferrous bicarbonates to ferric hydroxides. Also the accumulated ferric hydroxide flocs aggravate growth of iron bacteria. The enlarging cross section of the Hopper chamber allows for gradual decrease of velocities in the upward flow which will reduce the risk of flushing away of bacteria, which might reduce the bacteriological action.

The water from the upward filter is allowed to spill slowly over wider spilling perimeter of the Hopper chamber, evenly distributing the water to the 150mm thick sand filter. The sand filter bed acts as a strainer since the small suspended solids are retained at the top of the layer. The filtered water is collected through perforated pipes at the bottom of the filter and collected at the outlet pipe. The overall retention time of the whole unit seems to be in the order of 5-8 minutes.

Cleaning operation

Special emphasis was placed in the whole design to ease and simplify the cleaning process at village level. It is estimated that a frequent cleaning process could be very easily carried out by two women within a period of 30-45 minutes. The cover slab consists of three cast segments of concrete lids weighing only 10kg each which is easily liftable with handles provided.

The blocked hopper chamber with iron bacteria flocs could be washed out by opening outlet provided after filling the chamber with water and raking the granite chips. The cleaning of the sand filter was simplified with the provision of a commercially available non-woven fabric where the filtered materials blocking the filter operation could be taken out. In general, use of the fabric had increased the filter run length and also has decreased the ripening period for the Schmutzdecke after a cleaning process.

Construction

Simplicity in construction is one of the major advantages of the Design. This consists of a concrete base slab, a Hopper chamber (concrete) a brick outer chamber wall and pre-cast concrete cover slabs. Since the formwork and the layout is a standard one the works could be handled very satisfactorily employing intermediate level technicians or skilled labourers, with very little training.

The type of work available does not need any special equipment and could be completed at village level without having a centralized precast unit, as in the case of FINNIDA circular type.

The per capita cost of construction of such a unit could be completed at a very low cost of US$0.4. (Table 1)

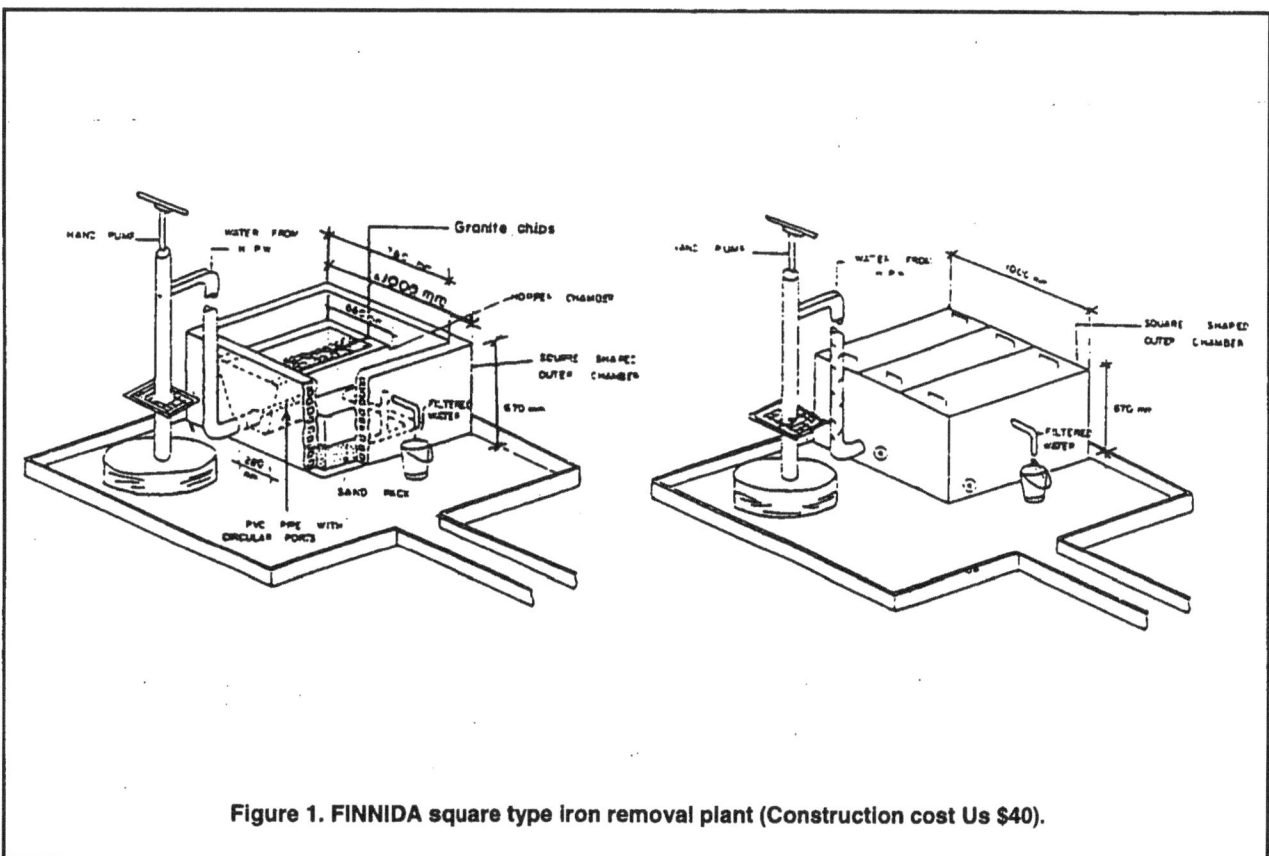

Figure 1. FINNIDA square type iron removal plant (Construction cost Us $40).

Table 1. Bill of Quantities for FINNIDA Square Type IRP

Item	Qty
Material	
20mm Aggregates	0.65cum.
6mm Aggregates	0.42 cum.
Sand	1.00 cum.
Mis. steel rods	22 kg
Bricks	200 Nos
Piping	5m
Labour	
Skilled labour	4 days
Labour	4 days

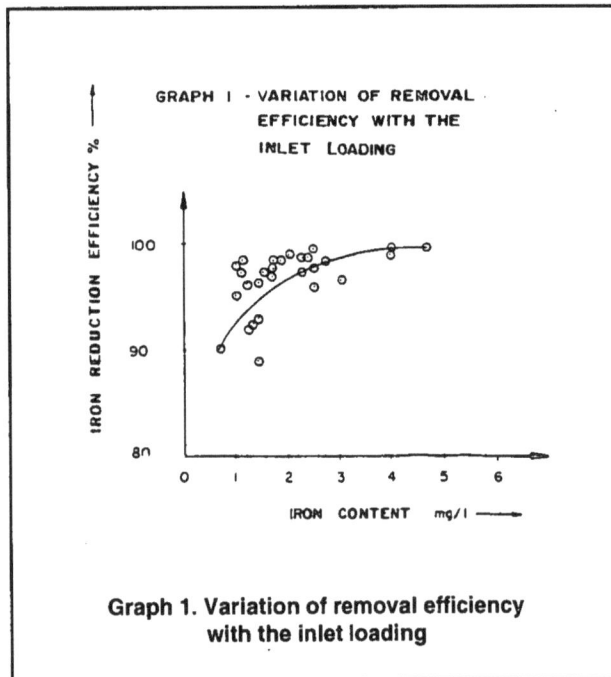

Graph 1. Variation of removal efficiency with the inlet loading

Monitoring of iron removal plants

Sixty handpump wells reported to have excessive iron in groundwater have been installed with both circular and square types over the last six years. This consists of 23 FINNIDA circular types installed in 1987-1988 and 37 FINNIDA square type installed since 1988.

The maintenance of these units had been completely handed over to the community. Intervention of the centralized maintenance unit had been completely stopped over the last six years limiting the project role to monitoring function only.

The detailed survey carried out on tall the sixty wells had revealed that out of the 23 Nos. of circular type, 37 Nos of square type IRPS 13 community. Intervention of the centralized maintenance unit had been completely stopped over the last six years limiting the project role to monitoring function only.

The detailed survey carried out on all the sixty wells had revealed 23 of circular type, 37 of square type IRPS 13 and 32 Nos respectively are in working order. The reason for the failures of the 13 Nos. of circular types IRPS are due to difficulties in frequent cleaning. On the other hand the balance survived in the operation are plants with low inlet loading which does not require frequent cleaning. The reasons given as causes of failures of five square type filters are unsuitable siting and lower yielding of handpump wells which are not dependent on iron removal technology.

The results of iron removal efficiency monitored over the last five years shows that the efficiency of the unit increases with increase of iron content of the inlet loading (Graph 1).

However, the following limitations are observed and needs further design development.

(a) Insect breeding under the concrete cover slab due to negligence of the consumers
(b) Damages caused on the edges of the cover slab due to frequent handling.

Conclusion

The FINNIDA square iron removal plant was found to be highly efficient in iron removal among taste sensitive communities. The technology of this unit seems to be more acceptable and adaptable at village level.

References

1. Ahmed F and Smith P.G. (1987) 'Design and Performance of a community type iron removal plant for hand pump tubewells.' *Journal of the Institution of the Water Engineers and Scientists* 41, 167-172.

2. Padmasiri J P & Attanayake M A S L (1991). 'Reduction of iron in groundwater using a low cost filter unit.' *Journal of the Geological Society of Sri Lanka* Vol. 3 Page 68-77.

3. Sri Lankan Standards (1983) Specifications for potable water 614, Bureau of Ceylon Standards, Sri Lanka 19 P.

Algae removal by roughing filter

J. Jayalath, J. Padmasiri, S. Kulasooriya, B. Jayawardena, W. Fonseka, and L. Wijesinghe

ANURADHAPURA SACRED CITY water supply scheme consists of an aerator, slow sand filters and a chlorination system. The source is a large irrigation tank, Tissa Wewa, situated close to the treatment plant. Frequent filter blockages and the bad odour of the filtered water are two major problems during this treatment. In an earlier study it was revealed that the predominant planktonic algae in the Tissa Wewa tank was Synedra sp. a pennate diatom which was present in association with a small number of cynobacteria and green algae. The accumulation of the silicified frustules of these diatoms blocks the filters and cause bad odour in filtered water. A pilot scale horizontal flow roughing filter was used to study the possibility of using a HRF to reduce the concentration of Synedra entering the slow sand filters. The reduction in the number of algae as well as the colour and turbidity were monitored. There was a marked reduction in the count of Synedra when pilot scale of HRF was used. Therefore a HRF can be used for pretreatment to reduce algae concentration in raw water. However, additional measures such as aeration should be employed to remove odour as this method does not guarantee odour removal.

Introduction

In an earlier study (Kulasooriya et. al. 1993) it was found that the raw water of the Anuradhapura Sacred city water supply scheme contains a high count of algae (Synendra which is a filter blocking type (Standard method for examination of water and wastewater, algae colour plates, 1985). High turbidity in raw water is another reason for frequent filter blockages in this scheme. However turbidity can be removed by using a horizontal flow roughing filter (HRF). Therefore an experimental model of a HRF was used to observe the change of algae count, colour and turbidity at different stages of the HRF.

Method

The experimental filter had three compartments (1m long) filled with filter media of three different sizes. The filter media used was granite. The filter was operated at a flow rate of 1.5 m/hr and the algae count, turbidity and colour were monitored at different points, namely the inlet, outlet and at two other points as shown in Figure 1.

Results and discussion

In the studies carried out in 1993, it has been reported that the highest count of algae at the intake was 191×10^3 and lowest was 1. By comparing the operational records of the existing slow sand filters and algal counts at the intake, the slow sand filters can be operated satisfactorily when the algal count is below 20×10^3 and turbidity below 15 FTU. According to the test results (Figure 2 and Figure 4),

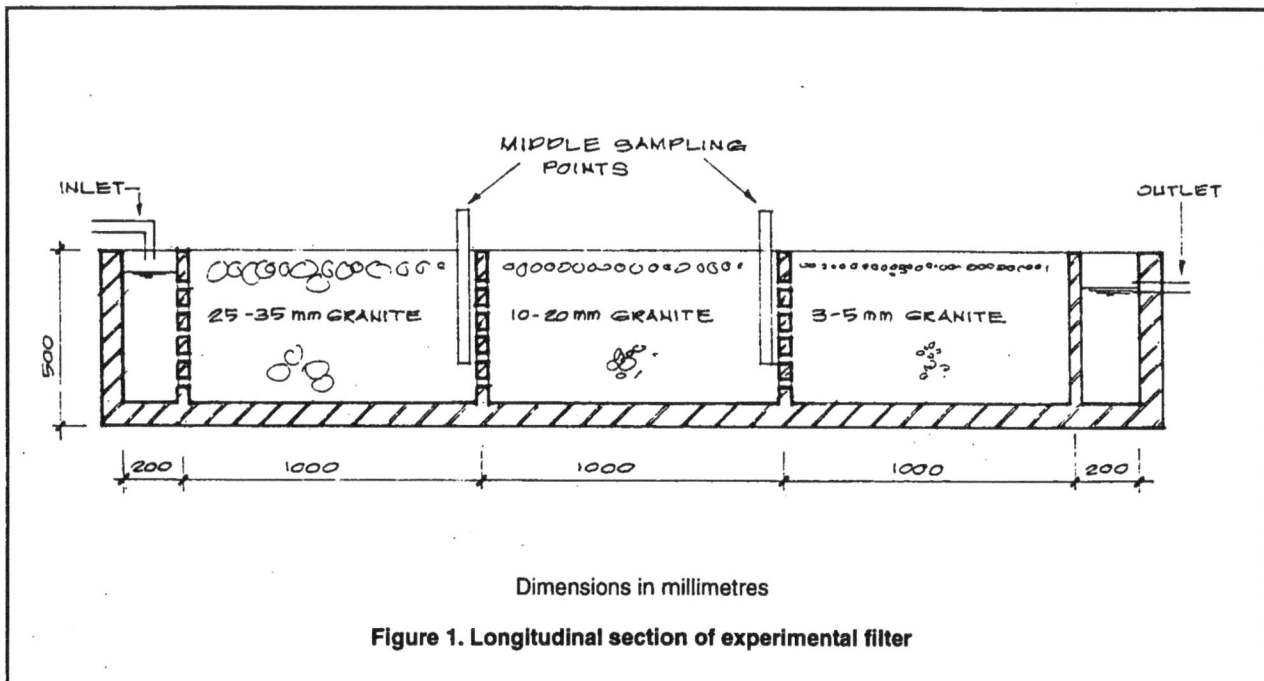

Dimensions in millimetres

Figure 1. Longitudinal section of experimental filter

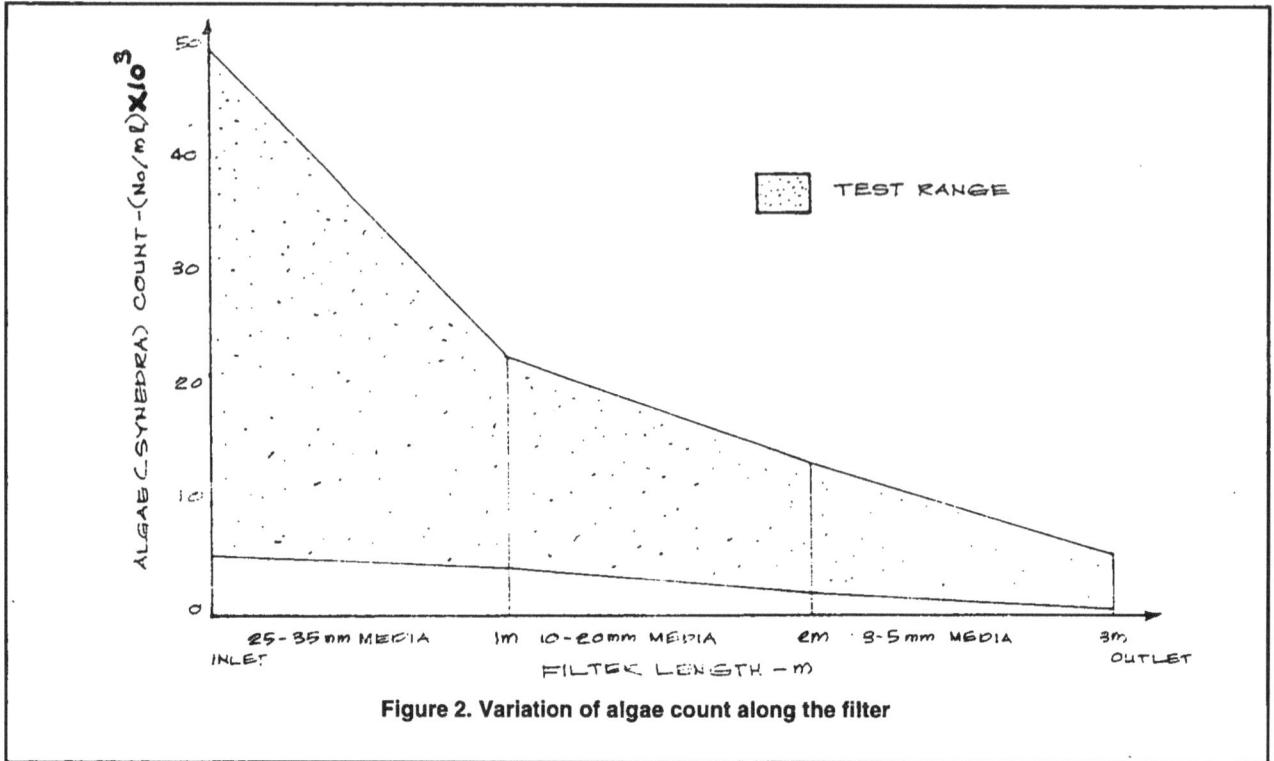

Figure 2. Variation of algae count along the filter

algal count and the turbidity are well below 20×10^3 and 15 respectively. Figure 2 shows that with a higher count of algae in raw water the rate of reduction in the count is high. The filter could be tested up to the maximum algal count of 55×10^3 and up to a turbidity count of 23 FTU. The results are within the safe limits for proper operation of slow sand filters. The length of the horizontal flow roughing filters used for water treatment is longer than the experimental filter. Therefore better results can be expected from the actual filters. In this method as the algae is collected inside the filter, formation of bad odour cannot be avoided. Therefore an odour removal system like aeration should be introduced. It was also noted that a remarkable reduction in colour also could be achieved by his process. The commonly used filter media in Sri Lanka is pebbles collected from river beds. Since granite is less expensive and readily available, granite was used in this experiment.

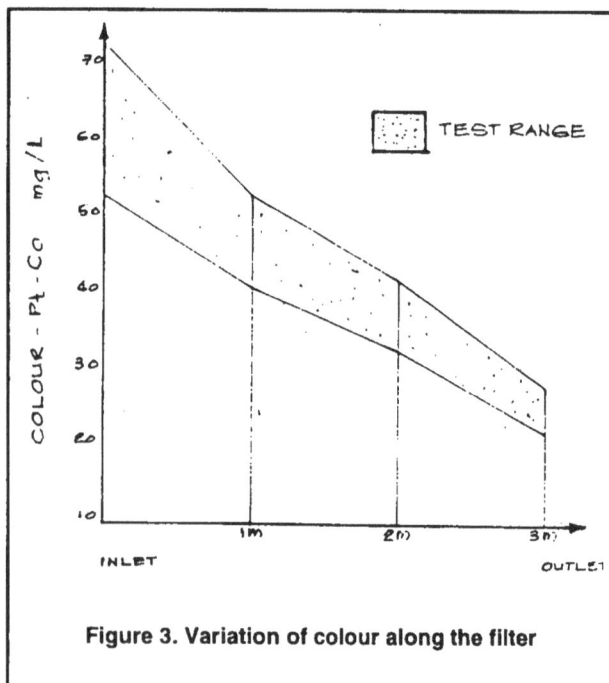

Figure 3. Variation of colour along the filter

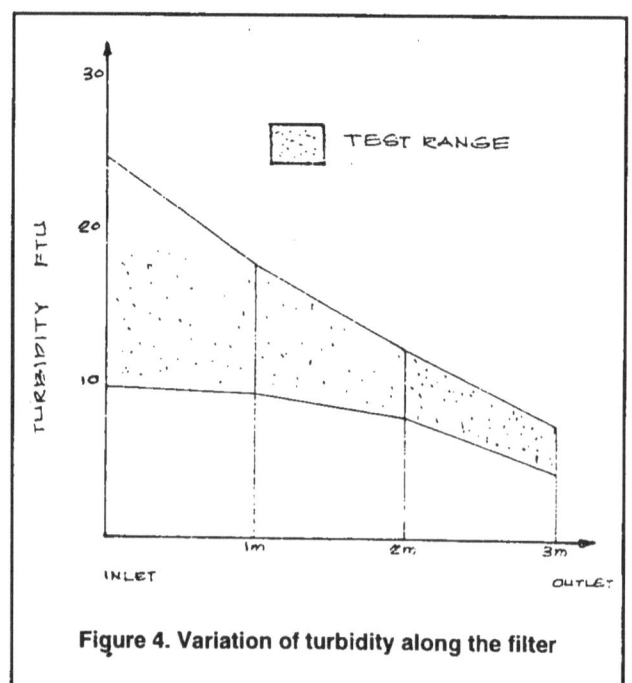

Figure 4. Variation of turbidity along the filter

Conclusion

Horizontal flow roughing filters with an odour removal system like aeration can be used for pretreatment when synedra is present in raw water.

Acknowledgement

The authors thank for the financial assistance given by the Natural Resources, Energy and Science Authority of Sri Lanka (Research Grant RG/92/B/06).

References

[1] Kulasooriya, S.A., Jayawardena, B.P.A., Padmasiri, J.P., Fonseka, W.S.C.A., some biological problems associated with water purification plant at the Tissa Wewa tank in Anuradhapura, Sri Lanka association for the advancement of science proceedings, part 1, pp 150, 1993.

[2] Wegelin Martin, *Horizontal flow roughing filtration (HRF) design, construction and operation manual*, IRCWD Report Np 06/86, 1986.

[3] *Standard methods for examination of water and wastewater*, APHA, AWWA and WPCF, Washington D.C. 16th Edition, Colour plates, 1985.

Defluoridation technology based on activated alumina

G. Karthikeyan, Mrs. S. Meenakshi, B.V. Apparao, Tamilnadu, South India

IN THIS PAPER the experiments related to the development of defluoridation unit at domestic level for a 3 mg/1 flouride water, using activated alumina are presented in detail. The design and other specifications of the defluoridation unit are given. Details regarding the field studies in a fluorosis - affected village near by Gandhigram are also discussed.

Materials and methods

Several grades of activated alumina with different particle sizes were first taken up for the study. After trial and error experiments, taking into consideration the defluoridation capacity of the material and the rate of flow of water through the bed of activated alumina, three different particle sizes were selected for further investigation with an aim to select the most suitable one among them for use in the defluoridation unit. The particle sizes of the three different grades of activated alumina studied, are: Grade I: >300 <355, Grade II: >150<300 and Grade III: >125<150 microns.

Column experiments were carried out in order to find out the variation of the rate of flow of water through the packed bed of activated alumina with a change in the height of the bed keeping the height of the water column above the bed as constant. These experiments were performed (with each of the three grades of alumina) at three different heights viz. 20, 25 and 30 cm of the activated alumina column. The diameter of the column used was 3.7 cm and perfectly dried alumina of each grade was used in the experiments.

The defluoridation capacities of the three grades of activated alumina were determined by column experiments, fixing the height of the column of alumina as 25 cm, the input water with 3 mg/1 fluoride and with an alkalinity of 432 mg/1 was actually selected from a fluorosis-affected village, Kolinjipatty, situated near by Gandhigram after carefully analysing various drinking water sources of that village for fluoride concentration by the fluoride electrode method (Fluoride electrode manual, 1977).

The concentration of fluoride in output water was monitored periodically and defluoridation experiments were continued till the output water fluoride did not exceed 1 mg/1 which is the tolerance limit as prescribed by the Bureau of Indian Standards. The defluoridation capacity of activated alumina was then calculated (S. Meenakshi, 1992).

Regeneration of exhausted alumina was carried out with two different regenerant solutions, viz., 2% sodium bydroxide and 2% hydrochloric acid.

Table 1.
Rate of flow of water through activated alumina

Grade	Particle size (microns)	Height of the column (cm)	Rate of flow of water (lit./hr.)
I	>300<355	20	4.6
		25	4.2
		30	3.7
II	>150<300	20	1.6
		25	1.3
		30	0.8
III	>125<150	20	0.9
		25	0.7
		30	0.5

Results and discussion

Results of the experiments on determination of rate of flow of water through the activated alumina column, at various heights of the column, at various heights of the column are given in Table 1.

These results indicate that there is a significant change in the rate of flow of water with a change in particle size as well as in the height of the column. The rate of flow is found to increase in the order with an increase in particle size because, the pore volume of the bed increases with increasing particle size. There is a decrease in rate of flow with increasin the height of the bed because the resistance offered by the bed itself itself increases with the increase in the height. Selection of a suitable height of the absorbent column is essential before we design any defluoridation unit because of two reasons:

(i) A minimum contact time is necessary between the fluoride and activated alumina particles.

(ii) On the other hand, the rate of flow should not be too low in which case there will be problem of acceptability of the design by the users.

The results of defluoridation capacities of the three grades of activated alumina are given in Table 2.

These results indicate that there is increase in defluoridation capacity in the order, grade I < grade II < grade III as expected. Out of the three grades, grade III

Table 2.
Defluoridation capacity of activated alumina

Grade (microns)	Particle size (microns)	Defluoridation capacity
I	>300<355	960
II	>150<300	1140
III	>125<150	1220

showed somewhat higher defluoridation capacity but with this grade the rate of flow of water is lowest. On the other hand, for the grade I the rate of flow of water is highest but it suffers from the disadvantage of having lowest defluoridation capacity among the three. As both the factors viz., defluoridation capacity and rate of flow of water are equally important (neither can be given preferential treatment over the other), it was felt reasonable to select the grade II which has good defluoridation capacity as well as good pore volume to give the necessary rate of flow of water.

The results of regeneration experiments showed that 2% HC1 is a better regenerant with a regeneration capacity of 2.2 x 10m^{-2} m^3 of alumina/Kg of HC1.

Based on the results obtained activated alumina grade II with the particle size > 150 < 300 microns was finally selected. Using this material the height of the column has to be finalised. Secondly, the quantity of the activated alumina to be taken in the unit is to be decided taking into account the cost factor also. Several trials were made with different quantities of activated alumina in stainless steel cylindrical columns of diffeerent dimensions. The most

optimum dimensions of the column were found to be as follows.

i)   Radius of the column = 5 cm

ii) Height of the column = 25 cm

With these dimensions of the column, the volume of the activated alumina bed comes to be 1.964 litres which accommodates about 2 kg of the material.

Using this stainless steel cylindrical column with the dimensions of 5cm radius and 25 cm height, attempts were then made to develop a suitable design of the unit with the provision for the input water reservoir at the top and the treated water collector at the bottom.

The radius of the column is such that a vertical three container design keeping the column in the middle, reservoir for input water at the top and the collector for treated water at the bottom is not easily acceptable by the users. Acceptability of any model by the users is the most important criterion for the success of any technology. Being conscious of this point, we surveyed the various water filters available in the market which are already accepted by the people. The objective of the survey is whether we can fit in our design into the already accepted model. After several attempts, it was possible to attach the column to the input reservoir of the existing filter itself. The design and the specifications of the defluoridation unit developed are shown in Fig.1.

The capacities of the various components of the unit are as follows.

Input reservoir capacity	=	12 lit
Column capacity	=	2 lit.
Capacity of treated water	=	13 lit

Details regarding defluoridation of water with this unit are as follows.

Rate of flow of water	=	3 lit/hr
Input water fluoride	=	3 mg/1
Output water fluoride	=	1 mg/1
Defluoridation capacity of the material	=	1140 mg F/Kg
Frequency of regeneration	=	3 months

At the rate of consumption of 12 litres per day, for a family of three to five, exclusively for drinking and cooking purposes, the material gets exhausted only after 95 days that is, approximately three months. The total cost of the defluoridation unit including activated alumina = Rs. 650.00. The recurring cost of defluoridation/annum for a family of three to five persons = Rs. 30.00.

Field trials

Field trials were carried out in a village Kolinjipatti, Nilakkottai block, Dindigul Anna District of Tamilnadu in South India. This unit was used by the members of the selected family continuously for a period of three months. Periodically (once in a week) the treated water samples

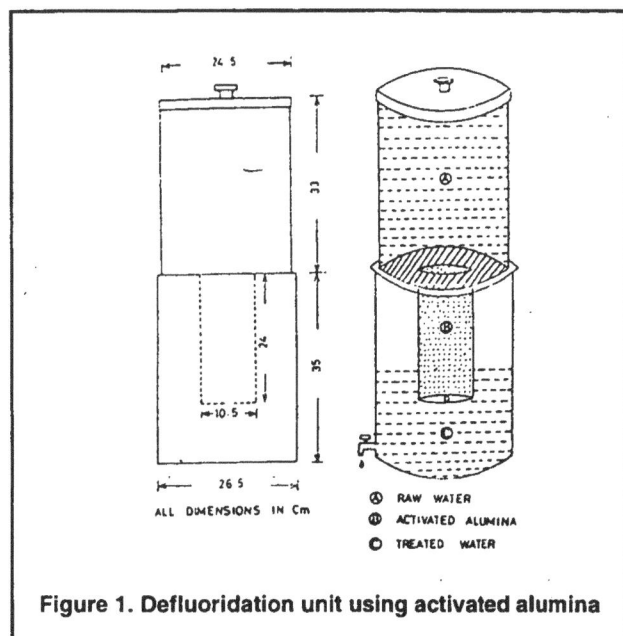

Figure 1. Defluoridation unit using activated alumina

Table 3.
Fluoride & pH levels of water samples during field trial

S.No.	Raw water		Treated water	
	Fluoride (mg/1)	pH	Fluoride (mg/1)	pH
1.	3.0	7.6	0.70	7.7
2.	3.0	7.7	0.72	7.9
3.	2.9	7.6	0.70	7.7
4.	3.0	7.8	0.75	8.0
5.	3.0	7.7	0.79	7.8
6.	3.0	7.5	0.81	7.8
7.	3.0	7.6	0.83	7.8
8.	3.0	7.7	0.85	8.0
9.	3.0	7.8	0.87	8.0
10.	3.0	7.6	0.88	7.9
11.	3.0	7.6	0.92	7.9
12.	3.0	7.7	0.93	7.9

were analysed for fluoride and pH in the laboratory and the results are given in Table 3.

Acceptability by the family

After just one week of using this defluoridation unit, the family got adapted to it and they have used the unit for a period of three months without any complaints. The fluorosis affected families in the village who, out of curiosity, visited the unit and got convinced of the ease of operation of the unit. The point that for the same cost of an ordinary water filter, they can get defluoridated water, impressed the people very much.

References

Fluoride Electrode Instruction manual, Orion Res.Inc., U.S.A., 1977.

Meenakshi A., Ph.D. Thesis, Gandhigram Rural Institute, Gandhigram, 1992.

Poly aluminium chloride as an alternative coagulant

Miss Sonu Malhotra, NEERI, Nagpur, India

ALUMINIUM SULPHATE COMMONLY known as alum is widely used as a coagulant in water treatment plants. Generally large sludge volumes are produced with alum which requires frequent desludging operations at the treatment plants causing increased wastage of water. There is also the possibility of aluminium carry over in water treated with alum. High levels of aluminium in potable water are reported to cause Alzheimer's disease, a form of senility. However at present there is no clear evidence to suggest a link between aluminium and Alzheimer's disease (Cole, 1990). Poly aluminium chloride (PAC) has been developed as an alternative coagulant for alum by an Indian manufacturer.

PAC hydrolyzes with great ease as compared to alum, emitting polyhydroxides with long molecular chains and greater electrical charge in the solution, thus contributing to maximize the physical action of the flocculation. Better coagulation is obtained with PAC as compared to alum at medium and high turbidity waters. Floc formation with PAC is quite rapid. The sludge produced by PAC is more compact than that produced by alum.

Preparation of coagulant solution

The working alum solution was freshly prepared by dissolving 10g of alum in one litre of distilled water. For making 1% solution of PAC the dilution of PAC was done with distilled water on daily basis.

Jar test equipment

All the laboratory tests were carried out using Phipps and Bird Multiple Stirring Device(Jar tester) equipped with stirring paddles and provisions for controlled mixing. The floc size and its settleability were observed with the illuminating device at the base of the apparatus.

Test conditions

Measured volumes (500 ± 10 mL) of samples were flocculated using the jar test apparatus in 600mL beakers. The beakers were placed in position in the jar tester. The motor of the stirrer paddles was started after addition of coagulant in each beaker simultaneously, rapid mix was maintained at 90 ± 10 rpm for 0.5 minutes followed by slow mix for high turbidity samples at 40± 5 rpm for 9.5 minutes. In case of low turbidity samples, after rapid mix and after addition of PAC or alum, slow mix was

done at 25 ± 5 rpm for 9.5 minutes. At the end of the stirring period, the beakers were removed slowly from the jar tester platform and the contents of the beakers were allowed to settle for 20 minutes. For each series, jar tests were repeated and average value of turbidity recorded to eliminate subjective errors. The criteria used for the evaluation of the efficiency of the coagulants were settled water turbidity and visual appearance of flocs.

Floc size index & flocculation

The floc size during jar tests were observed visually and recorded arbitrarily as per the following classification:

Floc Size Index (FSI)	Indication
1	No flocculation
2	Pin Point floc
3	Small floc
4	Large floc
5	Lump floc

The above classification was made by the author throughout the sets and observations were tabulated for each set of jar test conducted for comparison of floc size for PAC and alum.

The comparative performance of alum and PAC at four turbidity levels viz. 150 NTU, 550 NTU, 800 NTU and 2200 NTU are tabulated in Tables 1 to 4 and summarized under Table 5. The results of sludge volume produced by alum and PAC are tabulated in Tables 6 and 7. The experiments on sludge volume were performed using Imhoff cones. Results of residual aluminium with alum and PAC are tabulated in Table 8. Aluminium analysis was performed colorimetrically using ECR method (Standard Methods, 1992)

Summary and conclusion

PAC is an effective and useful substitute for solid alum which is conventionally used as a coagulant in most of the water treatment plants in India. It can cause rapid coagulation of water at different turbidities, produces less sludge and leaves less residual aluminium.

Table 1. Comparative performance of alum and PAC

Source : Wainganga river
Turbidity : 150 NTU

Alum, mg/L	0	5	10	15	20	30	40
Residual Turbidity,NTU	148	102	16	13	10	6	4
Reduction %	0	31.1	89.2	91.2	93.2	95.9	97.3
FSI	1	2	3	3	4	4	4
pH	8.2	8.0	7.8	7.6	7.5	7.4	7.2
Alkalinity mg/L as $CaCO_3$	168	164	160	159	158	152	148

PAC, mg/L	0	4	8	12	16	20	30
Residual Turbidity,NTU	152	106	33	18	13	11	4
Reduction %	0	30.3	78.3	88.1	91.4	92.8	97.4
FSI	1	2	3	4	5	5	5
pH	8.2	8.2	8.2	8.2	8.1	8.1	8.1
Alkalinity mg/L as $CaCO_3$	168	166	165	164	162	160	156

Table 3. Comparative performance of alum and PAC

Source: Narmada river
Test water turbidity: 800 NTU

Alum, mg/L	0	10	20	40	60	80
Residual Turbidity,NTU	770	294	93	16	8	5
Reduction %	0	61.8	87.9	97.9	98.9	99.3
FSI	1	2	3	4	4	4
pH	8.0	7.9	7.5	7.4	7.2	7.1
Alkalinity mg/L as $CaCO_3$	88	85	80	78	76	74

PAC, mg/L	0	5	10	20	30	40
Residual Turbidity,NTU	780	331	79	17	9	5
Reduction %	0	57.6	89.9	97.8	98.8	99.3
FSI	1	3	4	5	5	5
pH	8.0	8.0	7.9	7.8	7.7	7.6
Alkalinity mg/L as $CaCO_3$	88	88	87	86	84	84

Table 2. Comparative performance of alum and PAC

Source : Narmada river
Turbidity : 550 NTU

Alum, mg/L	0	10	20	30	40	50
Residual Turbidity,NTU	570	75	14	10	5	5
Reduction %	0	86.8	97.5	98.2	99.1	99.1
FSI	1	2	3	4	4	4
pH	7.7	7.3	7.2	7.1	7.0	6.9
Alkalinity mg/L as $CaCO_3$	85	84	82	81	80	78

PAC, mg/L	0	5	10	20	30	40
Residual Turbidity,NTU	565	129	44	24	10	4
Reduction %	0	77.2	92.2	95.7	98.2	99.2
FSI	1	3	4	4	5	5
pH	7.7	7.7	7.6	7.5	7.4	7.3
Alkalinity mg/L as $CaCO_3$	85	84	84	82	82	82

Table 4. Comparative performance of alum and PAC

Source: Kanhan river
Test water turbidity :2200 NTU

Alum, mg/L	0	20	30	40	50	60	70
Residual Turbidity,NTU	2250	80	29	28	17	9	4
Reduction%	0	96.4	98.7	98.7	99.2	99.6	99.8
FSI	1	3	4	4	4	4	4
pH	7.9	7.6	7.5	7.5	7.4	7.4	7.3
Alkalinity mg/L as $CaCO_3$	120	98	85	80	75	70	65

PAC, mg/L	0	10	20	30	40	50
Residual Turbidity,NTU	2270	200	29	8	5	5
Reduction %	0	91.2	98.7	99.6	99.8	99.8
FSI	1	3	4	5	5	5
pH	7.9	7.9	7.7	7.7	7.6	7.6
Alkalinity mg/L as $CaCO_3$	120	110	105	100	95	90

Table 5. Comparative dose to bring down the turbidity to 5 NTU

S. No.	Turbidity NTU	Alum mg/L	PAC mg/L	% PAC Consumption against % Alum Consumption
1.	150	35	25	71.4
2.	550	40	35	87.5
3.	800	80	40	50.0
4.	2200	65	40	61.5

Table 6. Residual Aluminium with PAC/alum

S. No.	Dose of Alum/PAC mg/L	Aluminium Concn. after treatment with, mg/L		Resultant pH	
		Alum	PAC	Alum	PAC
1.	0	0.05	0.05	8.0	8.0
2.	200	0.10	0.05	6.5	7.3
3.	300	>0.80	0.05	5.0	7.1
4.	400	>0.80	0.05	4.8	7.0

References

1) Cole, H; *Asian Water and Sewage*, p.53, 1990

2) *Standard Methods for the examination of water and waste water*, 18th edition, published jointly by American Public Health Association, American Water Works Association, Water Environment Federation, p.3-44, 1992

Table 7. Sludge volume produced by alum and PAC.

Raw water turbidity - 2010 NTU, pH -8.0, volume - 1 L
Source : Narmada river

S. No.	Time (min.)	Sludge Volume, mL	
		Alum dose 80 mg/L	PAC dose 50 mg/L
1.	0	0	0
2.	5	35	30
3.	10	27	21
4.	15	24	15
5.	20	20	16
6.	30	19	15
7.	45	17	14
8.	60	14	12
9.	90	14	12
10.	120	13	11
11.	150	13	10.5
12.	180	12.5	10
13.	240	11	9.5
14.	300	10.5	9.5
15.	1260	9	9.0
16.	1440	8.5	8.5

Table 8. Sludge volume produced by alum and PAC

Raw water turbidity - 1250 NTU, pH - 8.1, volume - 1 L
Source: Solapur water works (raw water)

S. No.	Time (min.)	Sludge Volume, mL	
		Alum dose 120 mg/L	PAC dose 80 mg/L
1.	0	0	0
2.	30	35	31
3.	60	31	27
4.	90	28	24
5.	120	26	21
6.	150	25	20
7.	180	23	20
8.	210	20	18

www.ingramcontent.com/pod-product-compliance
Lightning Source LLC
Chambersburg PA
CBHW080901030426
42336CB00016B/2976